Agriculture Issues and Policies

Additional books in this series can be found on Nova's website under the Series tab.

Additional e-books in this series can be found on Nova's website under the e-book tab.

AGRICULTURE ISSUES AND POLICIES

CROP GENETIC RESOURCES

CLIMATE ISSUES, ECONOMICS AND POLICY

SEAN W. PRICE
EDITOR

nova publishers
New York

NOTICE TO THE READER

Library of Congress Cataloging-in-Publication Data

ISBN: 978-1-63483-135-2

Published by Nova Science Publishers, Inc. † New York

CONTENTS

Preface		**vii**
Chapter 1	Using Crop Genetic Resources to Help Agriculture Adapt to Climate Change: Economics and Policy *Paul W. Heisey and Kelly Day Rubenstein*	**1**
Chapter 2	Crop Genetic Resources: An Economic Appraisal *Kelly Day Rubenstein, Paul Heisey, Robbin Shoemaker, John Sullivan and George Frisvold*	**33**
Chapter 3	The Seed Industry in U.S. Agriculture: An Exploration of Data and Information on Crop Seed Markets, Regulation, Industry Structure, and Research and Development *Jorge Fernandez-Cornejo*	**85**
Index		**241**

PREFACE

Climate change poses significant risks to future crop productivity as temperatures rise, rainfall patterns become more variable, and pest and disease pressures increase. The use of crop genetic resources to develop varieties more tolerant to rapidly changing environmental conditions will be an important part of agricultural adaptation to climate change. Finding new genetic traits that can facilitate adaptation—and incorporating them into commercially successful varieties—is time-consuming, expensive, and technically difficult. This book reviews the types of genetic resources, the ways they have been used, and how they might be used in the future. The book also discusses economic, scientific, and institutional factors that will determine the extent of genetic resource use and the benefits it might bring to climate change adaptation. Moreover, this title evaluates economic and institutional factors influencing the flow of genetic resources, including international agreements, and their significance for agricultural research and development in the United States. Finally, the book examines the composition of U.S. and international seed markets, regulations affecting agricultural seeds, the structure and evolution of the seed industry, and trends in private and public R&D in plant breeding.

In: Crop Genetic Resources
Editor: Sean W. Price
ISBN: 978-1-63483-135-2

Chapter 1

USING CROP GENETIC RESOURCES TO HELP AGRICULTURE ADAPT TO CLIMATE CHANGE: ECONOMICS AND POLICY*

Paul W. Heisey and Kelly Day Rubenstein

ABSTRACT

Climate change poses significant risks to future crop productivity as temperatures rise, rainfall patterns become more variable, and pest and disease pressures increase. The use of crop genetic resources to develop varieties more tolerant to rapidly changing environmental conditions will be an important part of agricultural adaptation to climate change. Finding new genetic traits that can facilitate adaptation—and incorporating them into commercially successful varieties—is time-consuming, expensive, and technically difficult. The public-goods characteristics of genetic resources can create obstacles to rewards for private research and development. Because of insufficient private incentives, public-sector investment in the use of genetic resources will help determine the agricultural sector's ability to maintain crop productivity, and for society as a whole, the potential benefits of public investment are large. The study authors find, however, that factors such as intellectual property rules for genetic resources and for research tools, or international

* This is an edited, reformatted and augmented version of Economic Information Bulletin Number 139, issued by the U.S. Department of Agriculture, Economic Research Service, April 2015.

agreements governing genetic resource exchange, have the potential both to promote and to hamper greater use of genetic resources for climate change adaptation.

Keywords: Crop genetic resources, crop germplasm, climate change, plant breeding, agricultural resources

What Is the Issue?

Climate change poses likely risks to future crop productivity as temperatures rise, rainfall patterns become more variable, and heat waves, drought, pests, and diseases increase. One strategy for helping farmers adapt to these changes is the development of crop varieties with better tolerance for increased stresses. Traits to boost crop adaptation, which may be found in genetic resources such as landraces (local varieties developed by farmers over many years) or in wild relatives of domestic crops, may be incorporated into varieties that gain wide commercial production. In light of their potential social and economic benefits, such resources have probably been underused. This report reviews the types of genetic resources, the ways they have been used, and how they might be used in the future. The report also discusses economic, scientific, and institutional factors that will determine the extent of genetic resource use and the benefits it might bring to climate change adaptation.

What Did the Study Find?

The authors first evaluated past and current use of crop genetic resources for stress adaptation and potential future demands on these resources. Their research led to the following conclusions:

- Empirical evidence from genetic resource use suggests that substantial economic benefits can follow from the collection, characterization, and use of genetic resources to improve crop resistance and adaptability.
- While the economic returns to genetic improvements in some U.S. crops have been substantial, contributions of the various stages of development—genetic resource discovery, conservation and use,

biological research, and breeding of new commercial cultivars—are not always easy to differentiate and assess.

- Climate change is likely to increase demand for new crop varieties with better resilience to stresses such as heat, drought, pests, and diseases. In recent years, demand for crop genetic resources from the U.S. National Plant Germplasm System (NPGS) has increased rapidly, even as the NPGS budget has fallen in real dollars.
- The use of new genetic traits in crops has concentrated more on tolerance to pests and diseases than to stresses like heat and drought that are also expected to increase with climate change.
- Recent scientific literature suggests genetically controlled traits responding to heat and drought may need to be considered together to have the greatest impact.
- Direct or easily traceable use of new genetic traits from sources such as landraces and wild relatives has been relatively rare for some crops but more common for others (e.g., potatoes and tomatoes).

Next, the study focused on technical, economic, and institutional and political factors that help explain the current pattern of genetic resource use and that are likely to affect future use. A summary follows of the results.

- Two kinds of technical change could reduce the costs of using genetic resources and thus increase their use for climate-change adaptation:
  - Improvements in genetic resource collection, conservation, characterization, and evaluation methods (e.g., through geographic information systems for predicting adaptation based on spatial information and DNA genetic marker analysis), and
  - Increased efficiency in incorporating valuable genetic traits into commercial crop varieties (e.g., through genetic markers linked to genes and gene segments that govern desired traits).
- Private firms may find it difficult to market and profit from the largely social benefits of genetic resource collection, conservation, and prebreeding activities. With insufficient private-sector incentives, the public sector is left to play a major role in optimal development of genetic traits to aid climate-change adaptation.
- Institutional factors such as international agreements and intellectual property rules can promote or hinder increased use of crop genetic resources to adapt to climate change, meaning that access to genetic

resources will be determined not only by supply and demand, but also by legal and political factors. This suggests the importance of considering unintended as well as intended consequences of such policies and agreements.

How Was the Study Conducted?

The authors reviewed the relevant scientific literature to determine how climate change may affect agricultural production, and thus the kinds of genetic traits that might be important for adaptation to climate change. They examined data from multiple sources to characterize past and current patterns of genetic resource use, including:

- Case studies of the use of particular genetic resources;
- Studies of the use of crop wild relatives for genetic improvement;
- Budget and distribution data for the National Plant Germplasm System (NPGS); and
- ERS analyses of data from USDA's Current Research Information System on U.S. public agricultural research, NPGS holdings from the Germplasm Resources Information Network, and information on traits used by the Germplasm Enhancement of Maize Project, a public-private collaborative effort.

The authors also reviewed the literature on the economics of genetic resource use and combined their observations into a supply/demand framework to analyze how genetic resource use for climate-change adaptation might increase over time. They also assessed existing intellectual property institutions and international agreements related to genetic resources.

Introduction

Genetic improvement has contributed markedly to sustaining growth in crop yields since the mid-20th century. However, climate change poses a significant risk to future crop productivity. The impacts of climate change on crop yields will be complex and will operate through many pathways. Crop varieties that are optimal for current growing environments are unlikely to be the best varieties for these environments in the future, as temperatures rise, rainfall patterns shift, and pest and disease pressures increase. Although current crop varieties can sometimes be grown in new locations as conditions change, new crop varieties, with attributes such as better tolerance not only to pests and diseases but also to heat and drought, could help agricultural producers adapt to climate change.[1] The development of crop varieties better suited to a changing climate will require increased efforts to incorporate traits from different genetic resources into new varieties.

Greater use of genetic diversity in plant breeding may be a relatively high-return, low-cost way of helping farmers adapt to climate change, but institutional and technical constraints might impede optimal exploitation of this option. Economic theory indicates that private investments in genetic resource use—conservation, evaluation, and eventual incorporation into crop varieties suitable for farmers' fields—depends on the expected returns. An inventor or institution can only profit from certain products during the research process. Unless a discovery can be sold, it will not be in a firm's self-interest to incur the costs of collecting, preserving, and characterizing crop genetic resources. There is near-consensus among both public and private-sector institutions[2] that the public sector has an important role to play in crucial stages of crop genetic conservation and improvement if the best results for society as a whole are to be achieved.

An additional economic assessment of the quest for economically valuable traits in genetic resources is called the economics of search. This report considers how the costs and benefits of such searches may be affected both by increasing demand, such as for greater tolerance to environmental stresses resulting from climate change, and by changes in technology, such as advances in geographic information systems, genomics, and related technologies.

Genetic Resources: What Are They, and How Are They Accessed and Used?

Germplasm Types

Germplasm (i.e., "seeds, plants, or plant parts that are useful in crop breeding, research, or conservation because of their genetic attributes" (National Research Council, 1993)), fall into four major types (table 1):

Table 1. Types of germplasm and other genetic materials used for crop improvement

Advanced germplasm	Recently developed varieties, obsolete varieties,* and other advanced breeding materials that are created with modern breeding techniques. These usually have higher frequencies of desirable genes than landraces or wild relatives.
Landraces	Varieties that were improved by farmers over many generations without modern breeding (generally requiring extensive efforts for use in a final variety).
Wild or weedy relatives	Undomesticated plants that share a common ancestry with a crop species (sometimes very difficult to incorporate in final varieties).
Genetic stocks	Germplasm that contains defined genetic variation (i.e., mutation), sometimes incorporated into an easy-to-use genetic base, which serves as a genetic research tool.
Cloned DNA sequences	"Transgenic" material from other organisms inserted into crops via molecular techniques. These differ from the other types of material because they cannot be used to regenerate an organism.

Source: USDA\Economic Research Service, based on Day Rubenstein et al. (2005) and National Research Council (1993).

*Obsolete varieties are older varieties no longer widely used but still useful for breeding.

- *Advanced ("elite") germplasm,* the advanced breeding materials developed in crop breeding programs.
- *Landraces, or farmer-developed varieties,* which often include more genetic diversity than advanced germplasm. Thus, they bring unique characteristics into the pool of available germ-plasm. However, it is more difficult to access and use landraces for crop improvement than to use advanced breeding materials.
- *Crop wild or weedy relatives,* which are generally even more difficult than landraces to use as a source of economically important traits in final varieties. This is mainly because they are genetically more distant from the crop of interest, and gene transfer by conventional crossing techniques ranges from problematic to impossible, making more radical techniques necessary.[3] This reduces their attractiveness as breeding material for some breeders, particularly those who work in the private sector, which tends to operate with shorter time horizons. Nonetheless, the use of crop wild relatives as a source of genes for developing new cultivars has been increasing over time (Guarino et al., 2011).

- *Genetic stocks,* which consist of material that contains defined genetic variations and are used as genetic research tools rather than for developing new varieties.

There is a fifth type of genetic resource that differs from the other four types because it cannot be used to regenerate an organism and usually is not held in genebanks:

- *Cloned DNA sequences,* or genetic material from other organisms incorporated into crops by molecular techniques (for example, a gene from the bacteria *Bacillus thuringiensis* used for resistance to insects.)

Collecting and Conserving Genetic Resources

Genetic resources can be conserved either "in situ" (in their natural setting) or "ex situ" (outside their natural setting) (Bretting and Duvick, 1997). Much of the world's plant genetic diversity occurs in situ. However, it usually needs to be collected before it can be used by breeders, who cannot readily access in situ material. The most important sources of in situ genetic resources can be found in regions where crops were originally domesticated. In situ resources can continue to evolve with climate conditions, expanding the range of available traits. However, climate change threatens the survival of some species in situ.

Ex situ conservation of some crops can store large amounts of genetic material as seeds at relatively low cost in terms of land required.[4] Breeders rely almost exclusively on ex situ materials. The world's genebanks presently hold more than 4 million distinct samples, consisting of crop varieties as well as other advanced materials, landraces, wild relatives, and genetic stocks.[5] However, only a small fraction of the world's plant genetic resources have been collected (Day Rubenstein et al., 2005).

The U.S. National Plant Germplasm System (NPGS) is the primary network that manages publicly held crop germplasm in the United States. This multi-institutional network is one of the largest national genebank systems in the world and is coordinated by USDA's Agricultural Research Service. The world's largest international system of genebanks is under the auspices of the Consultative Group on International Agricultural Research (CGIAR).[6]

The NPGS has extensive holdings, and it distributes germplasm to researchers, breeders, and educators free of charge and serves a large international scientific community. Table 2 shows NPGS holdings for the five U.S. crops with the largest gross value of commercial production, divided into the first four categories listed in table 1: (1) advanced materials, (2) landraces, (3) wild and weedy relatives, and (4) genetic stocks. The "improvement status" for NPGS accessions is sometimes unknown or simply unspecified. Wild relatives, for example, differ in the degree to which they are related to the crop of interest. The ease of trait transfer (and resulting improvement status) depends, in part, on this degree of relatedness.[7]

Table 2. Distribution of National Plant Germplasm System accessions among major U.S. field crops by germplasm types

	Crop				
Germplasm type	Wheat	Corn	Soybeans	Cotton	Alfalfa
Advanced*	19,634	9,120	18,077	1,362	1,452
Landrace	28,206	4,621	84	31	725
Wild relatives	7,326	275	2,113	1,890	1,105
Genetic stocks	4,103	7,511	828	0	2
Uncertain or unspecified	4,612	56	332	3,504	450

Source: USDA\Economic Research Service analysis of Germplasm Resources Information Network data.

With the exception of wild relatives, all accessions are listed by improvement status within the primary gene pool as defined by the Germplasm Resources Information Network (GRIN). Wild relatives may differ in their degree of relatedness to the crop of interest.

*Improvement status of advanced germplasm in GRIN includes an assessment of "breeding material," "cultivar," and "cultivated material."

For many of the major crops listed in table 2, the largest number of holdings consists of advanced material. Genetic stocks, used more for genetic research than directly for crop breeding, appear to be important for corn, wheat, and to some extent, soybeans. Landrace accessions are relatively common, particularly for wheat and corn. For these two crops in particular, the numbers of wild relative accessions appear small compared with landrace accessions. The number of accessions alone does not indicate critical gaps in NPGS holdings, but accessions can supplement the insights of genetic resource specialists. National working groups of specialists familiar with a given crop, called Crop Germplasm Committees, identify gaps in the collections and develop proposals for filling those gaps.[8] At the international level, the Global Crop Diversity Trust, an independent international organization established in 2005, addresses many issues concerning genetic resources, including strategies for collection, conservation, and use of crop wild relatives (Khoury et al., 2010; Dempewolf et al., 2014).[9]

Improving Crop Performance Under Climate-Change Conditions

Crops need tolerance to various types of stress.

Crop genetic resources can be used to increase yield by increasing resistance or tolerance to different types of stresses. Although genetic resources can also contribute to increased yield by raising physiological yield potential, we focus on stress tolerance here because increased plant stress is likely to accompany climate change. The first type of stress is "biotic," that is, due to the deleterious actions of pests or pathogens. The second type is "abiotic," or stress from the environment, such as extremes of moisture or temperature. As table 3 indicates, the current amount of research and knowledge about how to adapt to both abiotic and biotic stresses is roughly inverse to the degree of certainty we have about their occurrence with climate change.

For example, rising temperatures are expected with climate change, yet genetic sources of heat tolerance in crop varieties are probably less developed than sources of tolerance to other kinds of stress. Changes in pest and disease pressure are also expected with climate change, and sources of such biotic stress tolerance are much better known. But it is much more difficult to predict the types and locations of climate-change-related alterations in pests and diseases.

Biotic Stress: Impacts of Pests and Plant Diseases

To date, research into genetic approaches to stress tolerance has been heavily focused on biotic stress; however, the relative impact of abiotic stresses due to climate change is likely to increase more than biotic stress.

Because germplasm providing resistance to pests and diseases has been easier to find and to incorporate, more crop varieties have been developed that resist insects or complement weed-management strategies. In the public sector, for example, over the past decade the number of projects focusing directly on genetic approaches to pest and disease tolerance for all crops was more than 20 times the number of projects focusing on heat tolerance and nearly 8 times the number of projects focusing on drought tolerance (Malcolm et al., 2012).[10] For major crops, privately produced seeds often offer pest and disease resistance. Private companies are beginning to promote some seeds for drought tolerance or to indicate that such varieties are in their research pipelines. But it is difficult to find evidence of private research for heat tolerance.

Table 3. Types of stress and genetic adaptation to climate change

Sources of stress	Will new crop traits be useful for adap- tation to climate change-related stress?	Scientific knowledge of traits—identification, incorporation, and screening*
Temperature (overall and extremes)	Very likely—global climate models pre- dict increasing temperatures under a variety of scenarios in many geographic regions	Less clear—fewest public-sector re-search projects, no stated products in private-sector research pipelines
Drought and excess moisture	Likely—greater unpredictability in precipitation is expected, but global climate models differ about geographic distribution of precipitation changes	Somewhat clear—some public research projects, recently released or private-sector products in the research pipeline.
Biotic stresses (pests and diseases)	Unclear—greater pressure likely, but extent and severity unknown	More established—many public research projects, considerable private-sector products

*These assessments might vary somewhat across crops. Source: USDA\Economic Research Service.

In 2000, end users of NPGS materials were asked about the traits they sought and the usefulness of the samples they received for improving resistance in 10 major crops over the period 1990- 1999 (Day Rubenstein et al., 2006). Of the traits respondents sought to evaluate in the germ-plasm samples, biotic resistance or tolerance was by far the most frequently cited—37 percent of samples—regardless of the type of germplasm. Resistance to abiotic stresses was sought in only 14 percent of samples.

Documented use of genetic resources—particularly landraces—by a private-public collaborative Germplasm Enhancement of Maize (GEM) project in the United States has also demonstrated a focus on biotic stress tolerance in corn, as well as on quality or yield characteristics, with little emphasis on abiotic stress tolerance. Of GEM-developed varieties with traits specified, three-quarters were developed for biotic resistance and less than 2 percent for abiotic tolerance.

Similarly, to date, crop wild relatives have been used most frequently as sources of pest- and disease-resistant germplasm. Far more instances have been recorded of crop wild relative use for these purposes than for enhancing tolerance to abiotic stress or improving other traits (yield, quality, and fertility restoration, for example; see table 4). Nonetheless, genetic resource and genomics experts believe crop wild relatives will be particularly valuable sources of traits for adaptation to climate change (Dempewolf et al., 2014).

Table 4. Global use of germplasm from crop wild relatives has concentrated on pest and disease resistance[1]

	Number of wild relatives contributing			
	Pest & disease resistance	Abiotic stress	All other traits*	Total number of contributed traits
Wheat	16	1	5	19
Corn	1	0	0	2
Soybeans	0	1	2	4
Tomato	10	2	2	55
Potato	6	0	0	12
Rice	7	3	2	12
All other crops	22	5	6	40
TOTAL	62	12	17	144

Source: USDA\Economic Research Service calculations based on Hajjar and Hodgkin (2007) and Maxted and Kell (2009). Calculations take into account the likely overlap in reported use from the two sources. Cotton and alfalfa—two field crops with large economic value in the United States— were not surveyed.

*Yield, quality, male sterility, fertility restoration, etc.

[1]Data from released varieties and advanced breeding materials.

Abiotic Stress: Environmental Impacts

The use of genetic resources for abiotic stress tolerance can have major impacts; however, identifying and developing these traits may be more challenging than for traits for resistance to biotic stresses.

Despite being the focus of less research, traits that confer tolerance to environmental stresses have had major impacts on increasing the production of some crops. For instance, beginning in the 1960s, a short-growth-habit characteristic (referring to size), derived from landraces in Asia, was combined with other desirable production characteristics to create the semidwarf rice and wheat varieties that were the basis of widespread "Green Revolutions" in those crops. Much of the yield gains from these semidwarf varieties came from greater innate yield potential. But some of the gains came from resistance to lodging (i.e., the collapse of the cereal stem when it can no longer support its own weight). Lodging is caused by a combination of plant characteristics and environmental stress from rain, wind, or hail. (Resistance to lodging also allowed plants to be fertilized at higher rates.)

More recently, complex genetic combinations providing tolerance to abiotic conditions—including delayed flowering to enable greater flexibility in adjusting to daylight at low latitudes, aluminum tolerance, and calcium-use efficiency—were major factors in the large-scale expansion of soybean production in Brazil. The varieties used included landraces from the Philippines and Peru, as well as advanced germplasm from Brazil and the United States. The aluminum and calcium characteristics also permitted deeper rooting and greater drought tolerance (Spehar, 1994).

In the near future, other genetic contributions to abiotic stress tolerance from a wide variety of sources may have large impacts. *Oryza rufipogon*, a wild relative of rice, has contributed traits that enable rice varieties to better survive prolonged submersion in deep-water or flooded rice-growing environments. Major seed companies operating in the United States have recently released several corn hybrids with greater tolerance to drought, using sources as divergent as nonplant organisms (e.g., bacteria) through genetic engineering, as well as elite corn germplasm using conventional approaches to breeding (i.e., rather than genetic engineering).

Notwithstanding these advances, abiotic stress tolerance in many cases is a complex trait characterized by the actions of multiple genes, so the assumption that a genebank accession is either "drought tolerant," for example, or "not drought tolerant" is an oversimplification. (Even in breeding for resistance against biotic stresses such as diseases and pests, the genetic controls for key traits sometimes become more complicated as breeders move away from resistance strategies based on single genes.) For example, a plant variety may be drought tolerant, but that drought tolerance may not hold under higher temperatures. Therefore, it may be more advantageous to study the genetic responses to individual stresses such as heat and drought simultaneously rather than separately (Mittler, 2006; Cairns et al., 2013).

The increasing likelihood of using multiple genes to achieve greater environmental resilience suggests that alternative sources of traits (e.g., crop wild relatives, nonplant organisms) are complementary rather than competitive. It is unclear what the changes in relative demand will be for traits from, say, crop wild relatives vis-à-vis traits from nonplant organisms. In part, the demand may depend on access to genetic resources. But it is likely that the demand for genetic resources of *all* types will increase.

PUBLIC AND PRIVATE INCENTIVES FOR GENETIC ENHANCEMENT

Economic returns to genetic enhancement have been significant.

Crop genetic improvement (i.e., the combination of genetic resources, biological research, and plant breeding) can result in new varieties and better genetic material. Large economic returns to genetic improvements have been demonstrated. Approximately half the yield gains in major U.S. cereal crops since the 1930s are attributed to genetic improvements (see Day Rubenstein et al., 2005, and sources cited therein).

To adapt varieties to climate change, scientists and breeders will continue to look for genetic resources with helpful traits. The ability to develop new varieties depends on several factors. These include breeders' and scientists' access to genetic traits of interest; their technical ability to incorporate these traits, first into breeding materials and then into commercial varieties; and the economic incentives that motivate their work. Economic incentives generally differ for scientists/breeders in the public sector from those they face in the private sector. Genetic resources—in the sense of the genetic information contained in seeds or plant cells—have "public goods" characteristics. Public goods have two classic characteristics: nonrivalry (use of a resource does not reduce the amount available to others) and nonexcludability (others cannot be prevented from using a resource once it is first made available). When seeds or other plant-propagating materials are purchased, the genetic content of the materials is purchased at the same time. As a result, genetic resources are easily transported and replicated. Unless legal limits exist and are honored, it is difficult for an individual country, firm, or farmer to exclude others from using the genetic information in seeds or other materials for propagating crops, and other users have little incentive to pay an entrepreneur for developing that information. With no exclusion and a resulting lack of compensation, the seller lacks incentive to develop improved varieties or germplasm populations that can be sold.11

Furthermore, the usefulness of particular genetic resources is highly uncertain, and time horizons for improving genetic resources are long. A basic economic argument underlying public policy related to genetic resources is that the private sector tends to underinvest in research incorporating these resources. The private sector generates returns by selling marketable items, such as "finished varieties" (e.g., the seeds farmers plant in fields) or the research tools used to create final varieties. There are few private incentives for conducting "upstream" research; that is, for adding new samples of genetic resources (also called accessions) to genebanks, screening accessions for desired traits, and incorporating genes from unimproved genetic resources into the advanced breeding materials used to create marketable final varieties. These activities are long term and risky, and the potential for protecting intellectual property is less the earlier the stage of development. For example, advances in basic biology and the scientific understanding of plant functions generally are not products that inventors can prevent other people from using. Thus, for the private sector, only a part of the results of the long process of technical improvement can be protected for sale (or "appropriated"). Therefore, private-sector investments in genetic resource conservation and use are likely to be below those that are best for society as a whole.

Germplasm Conservation Benefits

The costs of conserving germplasm are small compared with the gains from genetic enhancement.

In 2012, The NPGS budget was approximately $47 million. Funding for the NPGS has been relatively stagnant over time. In real terms, agency funding peaked in 2003, at approximately $53 million in 2012 dollars (fig. 1). While direct comparisons between costs of a genebank and its benefits are not possible, for context, we note that U.S. farmers paid $20.3 billion for seed in 2012 (USDA\National Agricultural Statistics Service, 2013). Thus, the costs of public ex situ plant conservation in the United States are a small fraction—under half of 1 percent—of the value of the eventual seed market. At the same time that budgets have decreased, demand for NPGS germplasm has reached historic highs (fig. 2).[12]

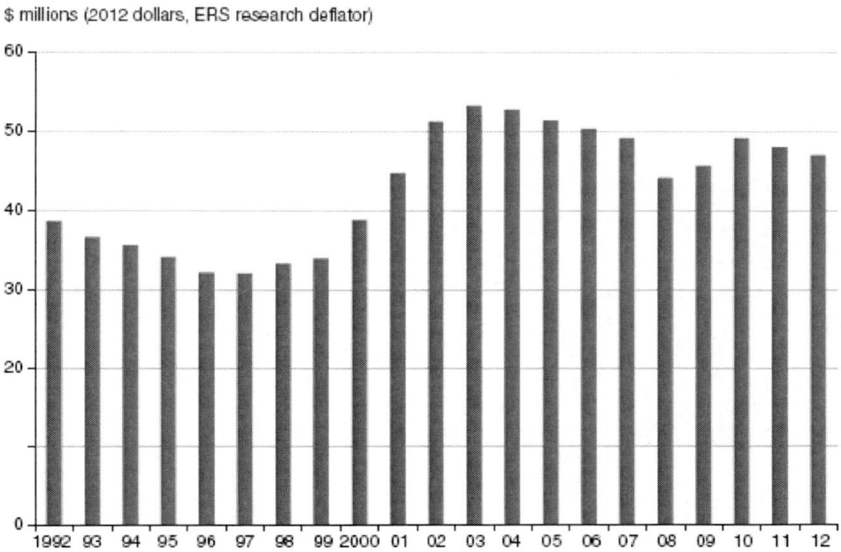

Source: Bretting (2013).
Note: In 2012, NPGS germplasm distribution was double its 2006 volume.

Figure 1. National Plant Germplasm System (NPGS) budget, 1992-2012.

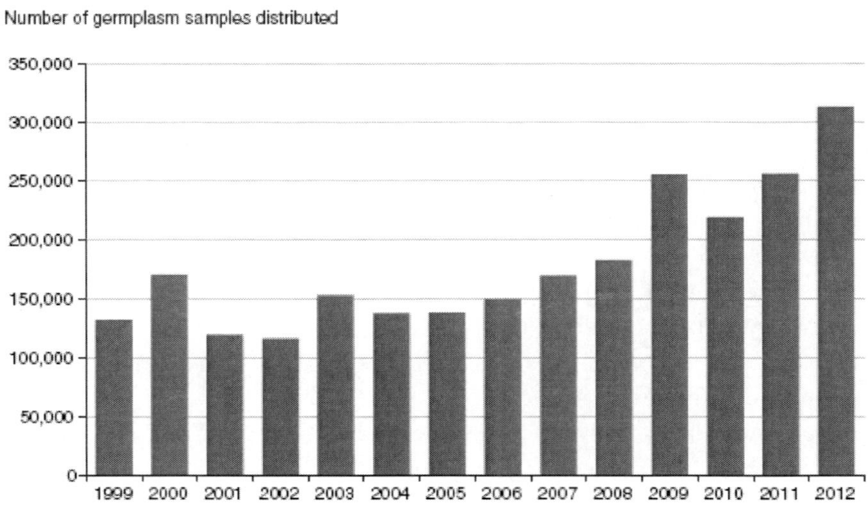

Source: Bretting (2013).
Note: In 2012, NPGS germplasm distribution was double its 2006 volume.

Figure 2. National Plant Germplasm System (NPGS) germplasm distribution, 1992-2012.

It is also not possible to make exact comparisons of the NPGS budget with other research expenditures on crop improvement, public or private, but estimates of related expenditures may help put these costs in perspective. In 2009, for example, total public-sector crop-related research investment in the United States was just over $1.7 billion (2012 dollars). Of this, just over $400 million was in the areas of "plant genome, genetics, and genetic mechanisms" and "plant biological efficiency and abiotic stress" (calculated from the USDA Current Research Information Systems database). In 2009, the private sector invested over $2.6 billion (2012 dollars) in crop-related research. Over $2 billion of this was in crop seed and biotechnology (Fuglie et al., 2011).[13]

Climate Change and Genetic Resource Use

Climate change may increase the benefits of genetic resource use; scientific advances may reduce the expense of identifying and incorporating helpful traits.

Because plant breeders usually breed advanced germplasm from their own program's gene pools, both public and private breeders incorporate only a small percentage of the material in genebanks into their programs. This is illustrated in the appendix to this report. Taking wheat as an example, the appendix provides estimates of the use of wheat landraces and wild relatives in wheat breeding programs. The available data, summarized in the appendix, suggest that use of both types of genetic resources has increased over time in global wheat breeding programs. At the same time, however, the data show that many of the programs do not use these kinds of genetic resources directly. Economic incentives favor incorporating traits from landraces and crop wild relatives only to the degree they are available through advanced materials, rather than incorporating them through direct use. Consequently, one economic question is whether the benefits from adding more accessions to gene-banks and incorporating them into breeding programs can justify the costs of these activities.

Economists have analyzed genetic resource acquisition and use through the application of search models (i.e., probabilistic and economic models that analyze which germplasm to search for, when to search for it, and when to stop searching for it).[14] A simple model is:

The expected benefits from investments in genetic resources

=

the probability of uncovering useful traits

x

the expected value of those traits.

Even if the probability of finding a useful trait is low, the total benefit might be high if the expected value of the trait is high. The optimal use of genetic resources will depend on the research costs, the value of the anticipated technological improvement, and the "discount rate" (essentially, the weight given to consuming in the present as opposed to consuming in the future). The type of materials also may affect the costs of searching among samples, and more exotic kinds of germplasm (e.g., landraces and wild relatives) may take much longer to add value to the final varieties. This may increase search costs and redirect efforts toward more easily transferred materials.

To illustrate the effects of climate change and of scientific advances on genetic resource use, we use an economic approach to examine financial investments in the identification and use of genetic resources. For now, we will set aside one of the major constraints identified earlier: insufficient private incentives. The simple assumptions we make concern only whether demand for genetic resources becomes greater with factors such as climate change and whether costs of using genetic resources fall with advances in technology.[15]

This analytical framework examines (1) the role of information (or discovering traits in germplasm); (2) the role of improved applied techniques (or incorporating traits in varieties); and (3) the need for climate-related traits. In figure 3, evaluating more genetic resource traits and incorporating them into commercial varieties is represented by moving to the right on the horizontal axis. Before considering the effects of climate change, the area under the "marginal benefit" line MB_0—at any point, say point A— measures the expected value of the benefits given that level of effort. The height of the line measures the expected value of analyzing and incorporating one more genetic resource. (The downward slope of the line illustrates the economic principle of diminishing returns.)[16] A simple version of this framework shows that each additional search increases the probability of finding the trait, but by a lesser amount as more genetic resources are searched. Increasing the scope of search or the resources devoted to incorporating useful traits into commercial varieties increases the probability of finding and incorporating useful traits, but by lower increments each time.

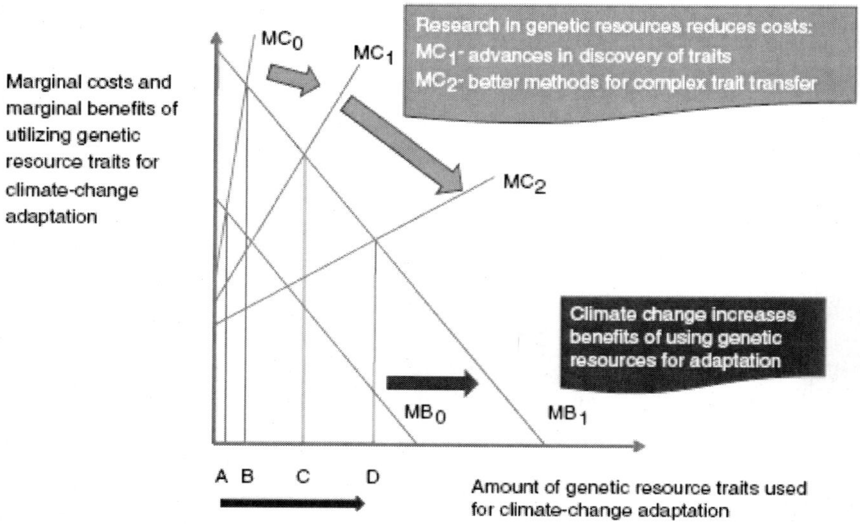

Source: USDA\Economic Research Service.

Figure 3. Climate change and research advances in crop genetics could increase the use of genetic resources.

The height of the "marginal cost" line, MC_0, shows the current cost of evaluating and incorporating an additional genetic resource. If new genetic resources need to be found, or if it becomes increasingly difficult to incorporate traits into successful commercial varieties, MC_0 would be upward-sloping, as depicted in figure 3.[17] In any scenario, as long as the expected marginal benefits of searching for and incorporating additional genetic traits are greater than the marginal costs, it makes sense to expend additional effort in search and incorporation activities, as this will increase expected total benefits, net of cost. If marginal costs are greater than marginal benefits, then research efforts should be reduced. The optimal level of effort is when the expected marginal benefits equal the marginal costs. In this first scenario, use of traits for climate change adaptation remains limited, at point A.

Climate change (observed or even anticipated) would increase the benefits from using genetic resources for better adapted varieties and would shift the marginal benefit line out to MB_1. This does not represent any change in the probabilities of finding useful traits from genetic resources or in the ease of incorporating these traits into useful varieties. Rather, the value of a trait like heat or drought tolerance, when found and incorporated into commercial varieties, would become greater as climate change leads to more problems in crop production. Genetic resource use would increase to point B.

If scientists could improve the process of trait discovery, the additional cost of finding a new trait would become lower as the probability of identifying a useful trait becomes higher.[18] In figure 3, as the marginal cost line becomes lower, MC_1, genetic resource use would move to point C. If, in addition to accelerated trait discovery, new and more efficient methods to transfer complex traits to finished varieties became available, costs would fall further to MC_2 and genetic resource use would move further, to point D.[19]

Costs of Genetic Resource Use: Current Evidence

The cost of searching for traits appears to be less than the value of the benefits of finding them, making it likely that genetic resource use for climate change adaptation will increase over time.

A few empirical studies have examined the economics of genebank acquisitions and use of gene-bank materials. An analysis of resistance to soybean cyst nematodes, a biotic stress, found that the expected marginal value of acquiring a new accession is anywhere from 36-61 times the marginal cost of such an acquisition, even though the probability of finding resistance in any particular acquisition is very low (Zohrabian et al., 2003). Another model of searching within an existing genetic resource collection applied empirical studies of resistance to Russian wheat aphid or septoria leaf blotch in wheat. Researchers concluded that just because some genebank accessions are not frequently incorporated into breeding programs does not mean that marginal accessions have low value (Gollin et al., 2000). When compared to the rather large benefits of genetic enhancement, the costs of genebank operation appear relatively small. Therefore, genetic resource conservation has probably been underfunded. Further, assuming genetic resource conservation is valuable, steady funding probably will be necessary as genetic resources must be conserved under controlled conditions and periodically regenerated to maintain their viability.

Scientific techniques related to genetic resource collection methods and to evaluation and incorporation of genetic resources into commercial varieties have benefited from technological improvements and are expected to benefit more over time. Increasingly sophisticated technologies that provide more information add usefulness to the germplasm acquired (i.e., shifting the marginal cost line downward in figure 3). For example, geographic information systems provide data on location (and by extension, climatic conditions) that may be particularly relevant for adaptation to changing temperature, precipitation, or day length.[20] Researchers have found that when genetic resources are accompanied by geospatial data, they are more useful (Day Rubenstein et al., 2006). New technologies (e.g., genetic markers, sequencing, and mapping, as well as genomic analysis) might lower both the costs of identifying useful traits and the costs of incorporating them into commercial varieties.[21] Such advances might reduce the costs and shorten the time lags associated with both trait searches and transfer of traits into advanced breeding materials, even when the source of the trait—such as a landrace or wild relative—has been traditionally more difficult for breeders to incorporate. The economic implications of using such techniques in the evaluation of genetic resources, along with an assessment of their costs, have not yet been addressed empirically.[22]

Returning to figure 3, the analysis suggests that, in addition to the possibility of insufficient private incentives, another reason for the relatively infrequent incorporation of genetic resources into breeding programs for heat or drought tolerance may be related to relatively high search costs and other costs to incorporate these traits compared with biotic resistance traits. The economic search models that are the basis of the analysis in this section may not capture the complexity of the search for genes that confer superior heat or drought tolerance. As noted earlier, all of the economic search models to date have focused on biotic stresses such as pest or disease resistance, and most of them have been based on expected probabilities of successful searches for single traits.[23]

IMPACT OF INSTITUTIONAL FACTORS ON GENETIC ENHANCEMENT

In addition to natural and scientific factors, institutional factors can also affect the utilization of genetic resources for adaptation to climate change (Noriega et al., 2013). These factors can be of particular concern because of the public goods characteristics of genetic resources. Institutional factors can shape access to and research on crop genetic resources and their commercial use. Institutional arrangements, such as international agreements and intellectual property rights, are generally intended to promote the use of crop genetic resources, but it can be difficult to design smoothly functioning and compatible legal arrangements and policies. Moreover, public investment will likely be necessary to promote the conservation, characterization, and enhancement of crop genetic resources, and global public investment requires even greater coordination than public investment within a given country.

Of particular importance are institutions that affect access to genetic resources that cross national boundaries. Like all countries, the United States depends in part on genetic material that is not already held in its own genebanks or found within its borders.[24] Wide access to genetic resources from multiple countries is needed.

Twenty-five years ago, germplasm was exchanged in a relatively open system (Noriega et al., 2013). Now, access to genetic material might be governed by one of several international agreements. The Convention on Biological Diversity ("the Convention") entered into force in 1992.[25] The Convention was developed under the auspices of the United Nations (U.N.) with the goal of conserving biodiversity. It recognizes nations' sovereign rights to their genetic resources, both in situ and ex situ. Exchange takes place primarily on a bilateral basis; countries have the power to negotiate their own terms. To collect genetic resources, the collector must have prior informed consent from the country holding the resource. Important provisions of the Convention include those addressing access and benefit sharing (ABS), which are intended to provide returns to the holders of genetic resources. In advance of guidelines that outlined implementation of ABS, several influential agreements promoted cooperation between countries that held genetic resources and pharmaceutical companies (Blum, 1993; Laird, 1993). The values of genetic resources for pharmaceuticals were assessed, and high returns were estimated (Mendelsohn and Balick, 1995).

However, the benefits of genetic resources, as seen by their holders, have often been more than the commercial values seen by industry (Laird and Wynberg, 2006). The use of genetic resources for agricultural varieties differs significantly from that for pharmaceutical products. Unlike drugs that can be used globally, agricultural varieties often need to be adapted for individual regions. As a result, crop genetic resources are rarely likely to yield the single, highly profitable developments that sometimes occur in pharmaceutical applications. Profit margins associated with products from medical research and development are generally higher than those for crop genetic improvement. For crops, overall benefits are likely to come from large numbers of small benefits. Because the returns to any particular genetic trait are likely to be limited, the incentive to conserve resources—particularly those in situ—may be inadequate.

On October 12, 2014, the Nagoya Protocol on Access to Genetic Resources and the Fair and Equitable Sharing of Benefits Arising from their Utilization entered into force. The Nagoya Protocol formalizes the Convention on Biological Diversity provisions for the sharing of benefits, provided that they conform to a nation's particular legislation.

A second agreement, the International Treaty on Plant Genetic Resources for Food and Agriculture (the International Treaty), addresses a subset of genetic resources from member countries for 64 food and forage crop species. The International Treaty has roots in the International Undertaking on Plant Genetic Resources for Food and Agriculture of 1983 (the International Undertaking), which provided for the free flow of genetic resources. The International Treaty was intended to encourage the exchange of the genetic resources found in the International Undertaking, while harmonizing exchange with the provisions of the Convention on Biological Diversity (Moore and Williams, 2011).

The incentive structures of the two agreements differ substantially as to how they motivate conservation of genetic resources. The Convention places the responsibility for negotiating genetic resources agreements primarily on individual countries, who determine the means of exchange (possibly including financial returns to the supplying institution). The International Treaty creates a more formal mechanism to facilitate the benefits of exchanging germplasm through a multilateral system, in which all members agree to share ex situ materials with each other via a single legal instrument called the Standard Material Transfer Agreement. The members also agree to share the benefits from the use of this material. The country or institution supplying a resource does not receive the financial returns from its germplasm; however, all members to the Transfer Agreement will benefit from access to and improved conservation of the genetic resource. The crops covered by the agreement include most of the major food crops important for food security, such as wheat and corn, but do not cover some others, including soybeans, peanuts, and tomatoes.

The United States has signed but not yet ratified the Convention on Biological Diversity and the International Treaty. Access to genetic resources by U.S. researchers and breeders can require extensive negotiations with different countries. In the case of the International Treaty, if the recipients, including the United States, accept germplasm from the Multilateral System, they also must accept its access and benefit-sharing provisions, which are triggered when materials are used in commercial products. In other words, if an accession from the International Treaty's Multilateral System is incorporated into a product that is sold commercially, some of the proceeds should be shared within the system according to a predetermined formula. Users of materials from the Multilateral System should not claim intellectual property rights that would limit access by others to that material in the form in which the users received it. Restricting access to a product (as with intellectual property protection) changes the benefit-sharing provisions. When a product using material from the Multilateral System is commercialized, recipients are encouraged to share 0.77 percent of the gross sales; however, if access to the product is limited through intellectual property rights, such a payment is mandatory (Moore and Williams, 2011). International genebanks of the CGIAR Centers follow the International Treaty's provisions. However, national genebanks either use the Standard Material Transfer Agreement when exchanging plant genetic resources with the United States or develop specific terms in bilateral negotiations with U.S. institutions for noncovered crops.

In terms of the number of accessions, the United States has a net positive balance of international exchanges of ex situ materials from the NPGS system. The NPGS distributes approximately six accessions to other countries for every accession it receives from other nations (Bretting, 2007).

The NPGS and other gene banks may want access to in situ genetic resources to add to existing collections. Recently, USDA has faced greater obstacles to collecting germplasm in other countries. This can be especially problematic in acquiring germplasm limited to a specific region. Many countries have not allowed USDA access, or the terms required are not acceptable for material entering the NPGS. As a result, USDA/ARS no longer requests permission for collecting expeditions (Williams, 2014).[26] Limitations on access to in situ resources reduce opportunities to collect materials such as crop wild relatives that may have traits useful for adaptation to climate change (Noriega et al., 2013).

Intellectual property (IP) rights also can affect genetic improvement directly through incentives that both foster innovation and limit access to new discoveries. IP rights offer limited-duration monopolies over inventions, offering creators opportunities to reap returns from their inventions. Historically, the United States has offered strong protections for intellectual property.[27] In the case of plant breeding, in the United States, IP restrictions applied to genes, such as patenting specific crop genetic resources,[28] are intended to create incentives for companies to invest in further improvement or distribution of the material in question (Noriega et al., 2013). However, the World Trade Organization, of which the United States is a member, has moved toward greater international harmonization of IP rights systems through the 1994 Agreement on Trade Related Aspects of Intellectual Property Rights

(TRIPS), which requires member countries to adopt minimum standards of IP protection. Nonetheless, different countries have maintained different levels of IP protection for crop genetic resources. One result may be that seed companies are less willing to invest in countries with lower levels of protection.

IP restrictions may also increase incentives to conduct related research. In addition to the patenting of improved genetic resources, some of the research tools needed for plant breeding can also be patented. Because many of the traits needed for climate change adaptation are expected to be complex and involve multiple genes, IP protections on the numerous research techniques used in plant breeding could have greater impacts than patents on individual genes.

Not all IP protection involves patents. The United States also uses Plant Variety Protection Certificates (PVPCs). Although PVPCs allow private crop breeders exclusive rights to multiply and market protected varieties, they offer weaker protections than patents because they include a research exemption that allows others to use the new variety for research purposes. Such protection is usually less contentious for international negotiations than stricter patent protections, and breeders make significant use of PVPCs (Fernandez-Cornejo, 2004). Unlike utility patents, PVPCs are not used for research tools because they pertain only to crop varieties.

Conclusion

Climate change is expected to place large demands for agricultural adaptation. Use of genetic resources to develop new crop varieties—more tolerant of both abiotic and biotic stresses—is likely to be an important component of agricultural adaptation to climate change. The usefulness of genetic resources in climate adaptation could be improved if:

- They are available (i.e., identified and preserved, probably in a combination of in situ and ex situ locations) and are widely accessible to plant breeders and researchers on acceptable terms;
- Information about the important characteristics of individual genetic resources are generated, recorded, and made widely available;
- Information concerning the genetics of complex traits are expanded; and

- Desirable characteristics are transferable to commercially competitive genetic materials. Private firms may have insufficient incentives to make such transfers, because the length of time required may conflict with the immediate need to produce varieties salable to farmers.

Economic analysis has demonstrated that although the costs of supporting these activities usually are relatively modest, potential benefits may be very large. Technological advances—either in more effectively screening genetic resources for desirable characteristics or in more rapidly incorporating these characteristics into commercially competitive germplasm—would likely increase the value of genetic resources. Furthermore, international agreements and intellectual property rights could possibly both increase access to genetic resources and incentives to pursue research. But as noted, such institutional constraints could also have the opposite effect, limiting international access to crop genetic resources or to the research tools needed to develop them.

The expanding options for climate-change adaptation would benefit from public support. Although social gains from genetic resource use may be large, the long time lags between initial research and commercially viable products highlight the financial riskiness of many of these activities. Furthermore, private appropriation of the social benefits from genetic resource use is often difficult. These concerns indicate that, in many cases, genetic resource collection, conservation, and prebreeding activities are likely to be primarily financed by the public sector, even where access to genetic resources is unlimited by institutional constraints.

Appendix: Contributions of Landraces and Crop Wild Relatives in Crop Breeding: Empirical Evidence for Wheat

Landraces and crop wild relatives have been found to make important contributions of traits used in crop improvements, and genetic resource managers have stated that diverse germplasm, particularly crop wild relatives, may become increasingly important sources of traits useful for crop adaptation to climate change (Dempewolf et al., 2014). At the same time, the use of landraces and wild relatives has been relatively uncommon in many crop breeding programs. These seemingly paradoxical statements are explained by the difficulty of incorporating traits from sources such as landraces or wild relatives and the time often required to transfer these traits into commercially successful germplasm. A few examples from wheat breeding illustrate this.

In 1981, Duvick (1984) surveyed leaders of breeding programs for five U.S. crops—cotton, soybean, wheat, sorghum, and corn—concerning the genetic diversity of those crops. Eighty-five percent of the wheat breeders said the breeding base in their program was broader than it was in 1970. *All* of the wheat breeders who said they had broadened their genetic base stated that one major source of new germplasm consisted of elite unadapted lines, in other words, advanced materials from other breeding programs in dissimilar growing environments. Twenty-five percent mentioned wheat land-races as a major source of broadening germplasm, and 15 percent noted “related species”—a category that would overlap with wild relatives of wheat—as an important source. Thus, some breeders had increased their use of landraces and wild relatives, but more were increasing genetic diversity only through access to advanced materials from other breeding programs whose genetic bases differed from those in their own programs.

Over time, the overall use of landraces and crop wild relatives in wheat breeding remained relatively small. In 1994, the International Maize and Wheat Improvement Center (CIMMYT) surveyed wheat breeders around the world, in both developing and high-income countries, concerning their use of different types of germplasm (Rejesus et al., 1996). Breeders were asked about their use of different types of materials in the breeding process (appendix table 1).

This study showed that globally, less than 8 percent of the entries in breeders’ crossing blocks were landraces. Landrace entries were relatively more common in developing countries, particularly in West Asia—the zone of origin for domesticated wheat—and nearby North Africa, both of which are separated out in the appendix table 1 summary.

Rejesus et al. also analyzed pedigrees of spring bread wheat releases in developing countries, and this analysis was updated by Smale et al. (2002). Between 1965 and 1997, fairly complete pedigree information was available for 1,162 releases of spring bread wheat cultivars by developing countries. Very few of these releases used a landrace, or a selection from a landrace, as a direct parent.[29] At the same time, the average number of *distinct* landraces in the pedigrees of spring bread wheat cultivars released by developing countries in the 1960s ranged between 20 and 30. The average number of *distinct* landraces in the pedigrees of releases in the mid-1990s ranged between 50 and 60. The annual rate of increase in unique landraces across all pedigrees was about one landrace per year. In other words, the diversity of unique landrace use was increasing over time not because most breeders were using landraces as direct parents, but because they had increasing access to advanced materials with more diverse genetic composition.

Regarding crop wild relatives, appendix table 1 shows that in 1994, wheat wild relatives were an even smaller part of wheat breeders' materials than were landraces. Indeed, the data are consistent with a scenario in which many, but not all, breeders were probably not using wild relatives. On the other hand, Khoury et al. (2010) state that over the past 20 years, the wild relatives of wheat have proved to be highly useful sources of resistance to both biotic and abiotic stresses in wheat. This implies that the use of wheat wild relatives in breeding programs continues to increase, albeit from a small base. The estimates for wheat reported in table 4 still represent only a tiny fraction of the total numbers of wheat advanced materials used or wheat cultivars released worldwide. This does not mean that these traits will not be economically important. Rather, it illustrates that in many breeding programs—given current technology—economic incentives favor incorporating traits from crop wild relatives only to the degree they are available through advanced materials, rather than incorporating them through direct use.

Appendix table 1. Type of parent materials in wheat breeders' crossing blocks, 1994

	Average percent of entries				
Region	Wild relatives	Landraces (local origin)	Landraces (foreign origin)	Landraces (TOTAL)	All other*
North Africa	0.0	14.0	6.5	20.5	79.5
West Asia	1.3	14.5	0.5	15.0	83.7
All developing countries	0.8	6.3	1.8	8.1	91.1
Former Soviet Union/East Europe	0.1	1.8	4.1	5.9	94.0
High income	0.9	1.8	4.9	6.7	92.4
World	0.8	4.5	3.0	7.5	91.7

Source: Rejesus et al. (1996).

*Includes own advanced materials, advanced materials from other countries, released varieties, international nurseries, and other unspecified materials.

REFERENCES

Association for Molecular Pathology v. Myriad Genetics Inc., No. 12-398 (June 13, 2013), United States Supreme Court.

Blum, E. 1993. "Making biodiversity profitable: A case study of the Merck/INBio agreement," *Environment* 35 (4):17-20; 38-45.

Bretting, P. 2013. "The USDA National Plant Germplasm System: 1999-2012." Presentation at the National Genetic Resource Advisory Council, March 5-6, 2013, Beltsville, MD.

Bretting, P.K. 2007. "The U.S. National Plant Germplasm System in an Era of Shifting International Norms for Germplasm Exchange," *Acta Horticulturae* (ISHS)760:55-60.

Bretting, P.K., and D.N. Duvick. 1997. "Dynamic Conservation of Plant Genetic Resources," *Advances in Agronomy* 61:1-51.

Cairns, J.E., J. Crossa, P.H. Zaidi, P. Grudloyma, C. Sanchez, J.L. Araus, S. Thaitad, D. Makumbi, C. Magorokosho, M. Bänziger, A. Menkir, S. Hearne, and G.N. Atlin. 2013. "Identification of Drought, Heat, and Combined Drought and Heat Tolerant Donors in Maize," *Crop Science* 53(4):1335-46.

Day Rubenstein, K., P. Heisey, R. Shoemaker, J. Sullivan, and G. Frisvold. 2005. *Crop Genetic Resources: An Economic Appraisal*. Washington, DC: U.S. Department of Agriculture, Economic Information Bulletin Number 2, 47 pp.

Day Rubenstein, K., M. Smale, and M.P. Widrlechner. 2006. "Demand for Genetic Resources and the U.S. National Plant Germplasm System," *Crop Science* 46(3):1021-31.

Dempewolf, H., R.J. Eastwood, L. Guarino, C.K. Khoury, J.V. Müller, and J. Toll. 2014. "Adapting Agriculture to Climate Change: A Global Initiative to Collect, Conserve, and Use Crop Wild Relatives," *Agroecology and Sustainable Food Systems* 38(4):369-77.

Duvick, D.N. 1984. "Genetic Diversity in Major Farm Crops on the Farm and in Reserve," *Economic Botany* 38(2):161-78.

Fernandez-Cornejo, J. 2004. *The Seed Industry in U.S. Agriculture*. Agricultural Information Bulletin No. 786, U.S. Department of Agriculture, Economic Research Service.

Fuglie, K., P. Heisey, J. King, C.E. Pray, K. Day Rubenstein, D. Schimmelpfennig, S.L. Wang, and R. Karmarkar-Deshmukh. 2011. *Research Investments and Market Structure in the Food Processing, Agricultural Input, and Biofuel Industries Worldwide*. Washington, DC:

U.S. Department of Agriculture, Economic Research Report Number 130, 147 pp.

Gollin, D., M. Smale, and B. Skovmand. 2000. "Searching an *Ex Situ* Collection of Wheat Genetic Resources," *American Journal of Agricultural Economics* 82(4):812-27.

Guarino, L., C. Khoury, and C. Fowler. 2011. "Adapting to climate change: the importance of ex situ conservation of crop genetic diversity," *Food Security and Climate Change in Dry Areas*, pp. 27-34.

Hajjar, R., and T. Hodgkin. 2007. "The Use of Wild Relatives in Crop Improvement: A Survey of Developments over the Last 20 Years," *Euphytica* 156(1-2):1-13.

Henry, R.J. 2014. "Genomic Strategies for Germplasm Characterization and the Development of Climate Resilient Crops," *Frontiers in Plant Science* 5(Article 68):1-4.

Khoury, C., B. Laliberté, and L. Guarino. 2010. "Trends in *Ex Situ* Conservation of Plant Genetic Resources: A Review of Global Crop and Regional Conservation Strategies," *Genetic Resources and Crop Evolution* 57(4):625-39.

Kolstad, C.D. 2000. *Environmental Economics*. New York: Oxford University Press.

Koo, B., and B.D. Wright. 2000. "The Optimal Timing of Evaluation of Genebank Accessions and the Effects of Biotechnology," *American Journal of Agricultural Economics* 82(4):797-811.

Laird, S.A. 1993. "Contracts for Biodiversity Prospecting." in: W.V. Reid, S.A. Laird, C.A. Meyer, R. Gamez, A. Sittenfeld, D.H. Janzen, M.A. Gollin, and C. Juma (eds.), *Biodiversity Prospecting: Using Genetic Resources for Sustainable Development*. World Resources Institute, Washington, DC.

Laird, S. A., and R. Wynberg. 2006. The Commercial Use of Biodiversity: An Update on Current Trends in Demand for Access to Genetic Resources and Benefit-sharing, and Industry Perspectives on ABS Policy and Implementation. UNEP/CBD/WG-ABS/4/INF/5, The Secretariat for the Convention on Biological Diversity, Montreal, Canada.

Malcolm, S., E. Marshall, M. Aillery, P. Heisey, M. Livingston, and K. Day Rubenstein. 2012. *Agricultural Adaptation to a Changing Climate: Economic and Environmental Implications Vary by U.S. Region*. Washington, DC.: U.S. Department of Agriculture, Economic Research Service, Economic Research Report Number 136, 84 pp.

Maxted, N., and S.P. Kell. 2009. *Establishment of a Global Network for the In Situ Conservation of Crop Wild Relatives: Status and Needs*. Rome: FAO Commission on Genetic Resources for Food and Agriculture, 266 pp.

Mendelsohn, R, and M.J. Balick. 1995. "The Value of Undiscovered Pharmaceuticals in Tropical Forests," *Economic Botany* 49(2):223-28.

Mittler, R. 2006. "Abiotic Stress, the Field Environment and Stress Combination," *Trends in Plant Science* 11(1):15-19.

Moore, G., and Williams, K.A. 2011. "Legal issues in plant germplasm collecting." In: *Collecting Plant Genetic Diversity: Technical Guidelines–2011 update*, L. Guarino, V. Ramanatha Rao, and E. Goldberg (eds.). Rome: Bioversity International.

National Research Council (NRC). 1993. *Managing Global Genetic Resources*. Washington, DC: National Academy Press.

Noriega, I. L., M. Halewood, R. Vernooy, E. Bertacchini, D. Gauchan, and E. Welch. 2013. "How Policies Affect the Use of Plant Genetic Resources: The Experience of the CGIAR," *Resources* 2(3):231-69.

Rausser, G.C., and A.A. Small. 2000. "Valuing Research Leads: Bioprospecting and the Conservation of Genetic Resources," *Journal of Political Economy* 108(1):173-206.

Rejesus, R.M., M. Smale, and M. Van Ginkel. 1996. "Wheat Breeders' Perspectives on Genetic Diversity and Germplasm Use: Findings from an International Survey," *Plant Varieties and Seeds* 9(3):129-47.

Simpson, R.D., R.A. Sedjo, and J.W. Reid. 1996. "Valuing Biodiversity for Use in Pharmaceutical Research," *Journal of Political Economy* 104(1):163-85.

Smale, M., M.P. Reynolds, M. Warburton, B. Skovmand, R. Trethowan, R.P. Singh, I. OrtizMonasterio, and J. Crossa. 2002. "Dimensions of Diversity in Modern Spring Bread Wheat in Developing Countries from 1965," *Crop Science* 42(6):1766-79.

Spehar, C.R. 1994. "Breeding Soybeans to the Low Latitudes of Brazilian Cerrados (Savannahs)," *Pesquisa Agropecuária Brasileira* 29(8):1.167-1.180.

USDA National Agricultural Statistics Service. *Agricultural Statistics 2012*. Washington, DC, U.S. Government Printing Office, p. IX-42. (Published 2013)

Williams, K.A. 2014. Personal correspondence, April 14.

Zohrabian, A., G. Traxler, S. Caudill, and M. Smale. 2003. "Valuing Pre-Commercial Genetic Resources: A Maximum Entropy Approach," *American Journal of Agricultural Economics* 85(2):429-36.

End Notes

[1] Climate change is also likely to lead to greater variability in weather. As a result, varieties that perform well under a wide range of conditions will be desirable.

[2] For example, the American Seed Trade Association, a trade organization of private seed companies, states "Federal investments to maintain . . . genetic resources are fundamental to the success of agriculture in the U.S." (See http://www. amseed.org/issues/state-federal/federal-key-issues/).

[3] Biotechnology can, and has, in some instances, reduced some of this difficulty by enlarging the range of "radical" techniques available to the plant breeder.

[4] Ex situ conservation in some cases (e.g., orchard plantings) can be expensive and require substantial land.

[5] "Other advanced materials" include breeding lines that have undergone significant genetic improvement but have not been released as a finished variety.

[6] Examples of member genebanks include the International Rice Genebank in the Philippines and the International Potato Center Genebank in Peru. Together, the CGIAR and the NPGS are the major distributors of plant germplasm internationally (Noriega et al., 2013).

[7] We follow the Germplasm Resources Information Network (GRIN) database of the NPGS in classifying species. For each crop, the scientific literature sometimes differs about how closely particular species are related to the crop of interest.

[8] For example, table 2 identifies a relatively large number of accessions of wheat wild relatives and landraces. Nonetheless, in 1996 the Wheat Crop Germplasm Committee identified as its first collection priority wheat wild relatives in regions in which they are native, ranging from Egypt across West Asia into the Turkic republics of Central Asia. The second priority was for landraces from countries such as Guatemala, where they had not previously been collected.

[9] The Global Crop Diversity Trust is funded by a diverse set of donors, including the United States through the U.S. Agency for International Development.

[10] A count of research projects was taken from USDA's Current Research Information System (CRIS) database.

[11] Some of the technical and institutional means that have been used to overcome this obstacle for private seed provision, in part, are alluded to below. For more detail, see Fernandez-Cornejo (2004).

[12] During the period 1990-99, about three-quarters of accessions distributed by NPGS were distributed domestically, with the rest distributed to countries outside the United States. Of accessions distributed domestically, just under a quarter went to private companies. The rest went mainly to nonprofit institutions, primarily USDA\ARS and universities (Day Rubenstein et al., 2006).

[13] There are often significant lags between private agricultural research investments and economic benefits, and even longer lags between public investments and economic benefits. Therefore, recent research expenditures are presented not as an indicator of likely current benefits, but for comparative purposes only.

[14] The first economic applications to genetic resources analyzed the question of whether potential pharmaceutical benefits from plants growing in the wild were great enough to induce private efforts to conserve the native habitats of those plants (Simpson et al., 1996). Economists later explored how information can affect the value of searches (Rausser and Small, 2000). In contrast to the search for traits with pharmaceutical value, traits with

agricultural value need to be incorporated into existing varieties in order not to lose traits from earlier selection and breeding.

[15] The search-model approach is not inconsistent with the fact that genetic resources have public goods characteristics. Using these assumptions, the same conclusions would hold even if a graphical representation of supply and demand of public goods (e.g., Kolstad, 2000, Chapter 5) is superimposed on figure 3. However this would complicate the discussion without altering the conclusions.

[16] This is similar to a simple graph illustrating returns to research in general; each additional dollar spent pushes the expected rate of return downwards.

[17] In many specific empirical situations, it might be difficult to characterize MC accurately. For example, it may not be linear—it could be a curve instead of a straight line. On the other hand, in the case of simply searching within an already-collected and defined set of genetic resources, MC could be flat rather than upward sloping. None of these modifications would change the conclusions about the effects of lowered costs for utilizing genetic resources.

[18] This discussion does not mean to imply that technological change in trait discovery or trait transfer has not already occurred. For example, the use of molecular markers in plant breeding has become an important practical tool over the last two decades. The discussion in the text refers to further technological advances.

[19] The analysis depends only on costs shifting downward, not on the specific geometry of the shift.

[20] Some crops, such as soybeans, are sensitive to the length of the day. These crops could move to different latitudes in response to changing temperatures or precipitation. If relocated, these crops' day length requirements would need to be adapted to their new location. (Day length itself does not change with climate change.)

[21] Genomics refers to investigations into the structure and function of large numbers of genes simultaneously. See Henry (2014) for an analysis of the potential contribution of genomics to characterizing crop wild relatives and to the development of climate resilient crops.

[22] Koo and Wright (2000) address, theoretically, search decisions such as when to begin a search, and assuming a desired trait is found, whether to develop a variety incorporating the trait.

[23] A further limitation is that current models assume that whether accession x possesses the trait in question is independent of whether accession y possesses the trait.

[24] Relatively few crop genetic resources originated in the United States. Native crops include sunflowers, strawberries, blueberries, cranberries, and pecans.

[25] The Convention on Biological Diversity is an international, legally binding treaty for all countries party to it. As of December 2013, all U.N. member States except the United States, Andorra, and South Sudan are parties to the treaty.

[26] There may be other instances of USDA scientists receiving material for uses other than those of the NPGS. Noriega et al. (2013) note that some GCIAR members also have had to reject materials because they are subject to too many restrictions.

[27] Article 1 of the U.S. Constitution: "The Congress shall have power ... To promote the progress of science and useful arts, by securing for limited times to authors and inventors the exclusive right to their respective writings and discoveries."

[28] In their natural state, genetic resources cannot be patented in the United States. (See Association for Molecular Pathology v. Myriad Genetics, 2013).

[29] One notably successful commercial variety that used a landrace selection as a parent was the Turkish cultivar "Gerek 79."

In: Crop Genetic Resources
Editor: Sean W. Price ISBN: 978-1-63483-135-2

Chapter 2

CROP GENETIC RESOURCES: AN ECONOMIC APPRAISAL*

Kelly Day Rubenstein, Paul Heisey, Robbin Shoemaker, John Sullivan and George Frisvold

ABSTRACT

Crop genetic resources are the basis of agricultural production, and significant economic benefits have resulted from their conservation and use. However, crop genetic resources are largely public goods, so private incentives for genetic resource conservation may fall short of achieving public objectives. Within the U.S. germplasm system, certain crop collections lack sufficient diversity to reduce vulnerability to pests and diseases. Many such genetic resources lie outside the United States. This chapter examines the role of genetic resources, genetic diversity, and efforts to value genetic resources. The report also evaluates economic and institutional factors influencing the flow of genetic resources, including international agreements, and their significance for agricultural research and development in the United States.

---

* This is an edited, reformatted and augmented version of Economic Information Bulletin Number 2, issued by the U.S. Department of Agriculture, Economic Research Service, May 2005.

SUMMARY

Half the yield gains in major U.S. cereal crops since the 1930s are attributed to genetic improvements. Demand for crops continues to grow, and environmental conditions change, so continued productivity growth—and the genetic diversity that helps sustain it—remains important. Genetic diversity can be conserved in farmers' fields, in ecosystems that contain wild relatives of cultivated varieties, and in national or international germplasm collections. It is difficult to determine the best mix of conservation strategies. Regardless, the use of genetic resources by one farmer or plant breeder does not preclude their use by another, so private incentives to sustain diverse genetic resources are low. This motivates public measures (and underlying research) to conserve genetic resources.

What Is the Issue?

Crop genetic resources are the basis from which all crop production stems. But habitat loss, the dominance of scientifically bred over farmer-developed varieties, and genetic uniformity are all threats to continued diversity. Plant breeders need diverse germplasm to sustain productivity growth. The U.S. system for genetic resource conservation may lack sufficient diversity to reduce some crops' vulnerability to pests and diseases. The genetic uniformity of many modern crop varieties has also raised concerns that crop yields and production will become more vulnerable to evolving pests and diseases. At the same time, genetic resource conservation is expensive, and both private incentives and public funding are limited.

Many sources of diverse genetic resources lie outside the United States. To slow or prevent loss of crop genetic diversity worldwide, international agreements have been designed to encourage preservation of genetic diversity and promote the exchange of germplasm. For example, the new International Treaty on Plant Genetic Resources for Food and Agriculture will govern the exchange of germplasm for crops like wheat, maize, and cotton. But implementation has been hampered by a lack of consensus among the treaty's parties on the value of particular genetic resources. Thus, many of the treaty's provisions, such as procedures for transferring germplasm, are still vague. U.S. policymakers and genetic resource managers will face new exchange terms and rules governing the sharing of benefits from commercialized products among the treaty's parties, so the time is right to examine of the costs and benefits of conserving genetic resources.

What Did the Study Find?

Since crop genetic resources are largely public goods, private returns to the holders of crop genetic resources are lower than their values to the world. Thus, private incentives for conservation are likely not sufficient to achieve a level of crop genetic diversity that is socially optimal. Significant economic benefits derive from conserving and using genetic resources. For example, a one-time, permanent yield increase from genetic improvements for five major U.S. crops has generated an estimated $8.1-billion gain in economic welfare worldwide. The estimated stream of benefits from genetic enhancement activities exceeds the cost of investments in genetic resource preservation and use. Consumers in both the developed and developing world have benefited from higher yields and lower world prices for food. Without continued genetic enhancement using diverse germplasm from both wild and modified sources, the gains in crop yields obtained over the past seven decades are not sustainable, and yields might eventually grow more slowly (or even decline). Agricultural production increasingly relies on "temporal diversity," changing varieties more frequently to maintain resistance to pests and diseases.

Three factors contribute to loss of genetic diversity—habitat loss, conversion from landraces (farmer-developed varieties) to scientifically bred varieties, and genetic uniformity in scientifically bred varieties. The loss of wild relatives occurs mainly through habitat conversion for agricultural use. Habitat loss is particularly problematic in developing countries, which often face greater pressures for wild land conversion than do developed countries. Crop genetic diversity also has diminished as landraces are displaced by scientifically developed varieties. Studies show that far less area is planted to landraces worldwide than a century ago. Finally, crop genetic diversity may decline with reductions in total numbers of varieties, concentration of area planted in a few favored varieties, or reductions in the "genetic distance" between these varieties. Thus far, yields for many major crops have been relatively stable as a result, at least in part, of frequent changes in modern varieties and breeders' continued access to diverse genetic resources.

This economic assessment suggests that crop genetic resources are essential to maintaining and improving agricultural productivity. However, a General Accounting Office (1997) study found that current conservation efforts may fall short of what scientists believe are necessary levels for future crop breeding needs, suggesting a role for public policy. Policy initiatives include broad-based programs of multilateral and bilateral financial assistance, stronger intellectual property rights, and international agreements for germplasm exchange. But institutional constraints may prevent these initiatives from achieving their stated goals.

How Was the Study Conducted?

This chapter examines the role of genetic resources and genetic diversity in agricultural production, and efforts to value genetic resources. From a review of published literature, the report addresses the value of genetic improvements over time and among regions of the world. Given the role of genetic diversity in minimizing pest and disease epidemics, the report explores how incentives for land conservation, the breeding process, and access to modern varieties can affect diversity in the field.

The report also evaluates economic and institutional factors influencing the flow of genetic resources—including international agreements such as the Convention on Biological Diversity and the International Treaty on Plant Genetic Resources for Food and Agriculture—and their significance for agricultural research and development in the United States. This chapter synthesizes existing literature to review three proposed policy tools to conserve plant genetic resources: (1) public investments in genetic resource preservation in their natural settings (*in situ* conservation) and of genetic resources saved in gene banks (*ex situ* conservation); (2) stronger intellectual property rights over genetic inventions, particularly in developing countries; and (3) agreements for transferring genetic materials among countries.

INTRODUCTION

Genetic resources provide the fundamental mechanics that enable plants to convert soil, water and sunlight into something of critical value to humans—food. Diverse genetic resources allow humans to select and breed plants and animals with desired characteristics, thus increasing agricultural productivity. U.S. agricultural productivity more than doubled over the last century (Ahearn et al., 1998), and much of this productivity increase came from rapidly rising crop yields. Half the yield gains in major U.S. cereal crops since the 1930s are attributed to genetic improvements (OTA, 1987). But demand for agricultural commodities continues to grow, and environmental conditions change, so continued productivity growth—and the genetic diversity that helps sustain it—remains important.

Genetic diversity can be conserved in the form of diverse cultivated varieties in farmers' fields, ecosystems that contain wild relatives of cultivated varieties, and/or germplasm collections that contain samples of wild and cultivated species. Each method is characterized by different costs and benefits, making it difficult to determine the optimal mix of conservation strategies. But each also shares a common feature. The use of genetic resources by one farmer or plant breeder does not generally preclude their use by another, so private incentives to hold and protect genetic resources are generally lower than their value to users as a group or society as a whole. This means that in the absence of appropriate public measures (and underlying research), private efforts to conserve genetic resources are likely to fall short of the conservation levels that are optimal for society.

Previous researchers have contributed to our knowledge about the use and conservation of genetic resources. The National Research Council published a detailed review of the National Plant Germplasm System that included extensive recommendations to improve the system (NRC, 1991). A second, related book presented a broader look at the management of genetic resources (NRC, 1993) and included chapters on economic value and ownership. However, economic methodology has evolved rapidly since this chapter was released, as have the policy instruments that are used to protect and exchange genetic resources.

The Food and Agriculture Organization of the United Nations developed a report based on studies submitted by member countries. *The State of the World's Plant Genetic Resources for Food and Agriculture* (1996b and 1998) was a useful snapshot of genetic resource conservation and technological methods, but provided minimal economic information such as incentive structures or policy tools. In 1997, the U.S. General Accounting Office presented a systematic analysis of the management of the U.S. national genebank system. Recently, the International Food Policy Research Institute published a set of research briefs focused on gene bank valuation. These last two reports focused only on gene banks, and not on all three genetic conservation options.

All these previous reports have been useful, but recent developments in the international exchange of genetic resources call for a concise and current summary of genetic resource conservation in an economic framework. This chapter focuses on our current understanding of the value of genetic resources, trends in genetic diversity (and the economic incentives that affect them), and recent strategies for protecting genetic resources (including the International Treaty on Plant Genetic Resources for Food and Agriculture, which entered into force in June 2004).

Origins of Crop Genetic Diversity

Human selection of plant varieties for desired traits (such as taste, pest resistance, or seed size) dates from the very beginnings of agriculture. For thousands of years, farmers have selected, saved, and replanted varieties of the crops that humans consume today. "Centers of diversity" developed where intraspecies diversity of crop varieties was particularly high. Most centers of diversity are found where crops were first domesticated, primarily in today's developing countries.

The pace of genetic improvement accelerated with the development of modern breeding techniques that facilitated selection of specific desirable traits. Breeders have crossed different parental material and selected traits to achieve high yields and improved quality for all types of crops. Breeders have also sought resistance to pests, diseases, drought, and other stress. In fact, resistance has become the primary goal of breeding for many crops.

DEFINITIONS

Biological diversity refers to the number, variety, and variability among plant, animal, and microorganism species and the ecological systems in which they live. Biological diversity can be defined at three levels. *Genetic diversity* refers to the different genes and variations generally found within a species. The variation among genes across different wheat varieties is an example. *Species diversity* refers to the variety and abundance of different species in a region. Finally, *ecosystem diversity* is exemplified by the variety of habitats, such as grasslands or wetlands, occurring within a region. The term biological diversity can refer to any or all of the three levels of diversity, but in this chapter we will focus particularly on genetic diversity in agricultural crops.

Crop genetic diversity can be conserved in its natural setting (i.e., *in situ*), or it can be collected and conserved outside its natural environment (i.e., *ex situ*). Within the context of crop genetic diversity, there are five basic kinds of genetic resources:

1. **Wild or weedy relatives** are plants that share a common ancestry with a crop species but that have not been domesticated. These can also be a source of resistance traits, but these traits may be difficult to incorporate in final varieties.
2. **Landraces** are varieties of crops improved by farmers over many generations without the use of modern breeding techniques. These varieties are generally very diverse within species, because each is adapted to a specific environment. Within a modern breeding program, they are sometimes used for resistance traits, and extensive efforts are generally required before their genes are usable in a final variety.
3. **Improved germplasm** is any plant material containing one or more traits of interest that has been incorporated by scientific selection or planned crossing.
4. **Advanced (or elite) germplasm** includes "cultivars," or cultivated varieties, suitable for planting by farmers, and advanced breeding material that breeders combine to produce new cultivars.
5. **Genetic stocks** are mutants or other germplasm with chromosomal abnormalities that may be used by plant breeders, often for sophisticated breeding and basic research.

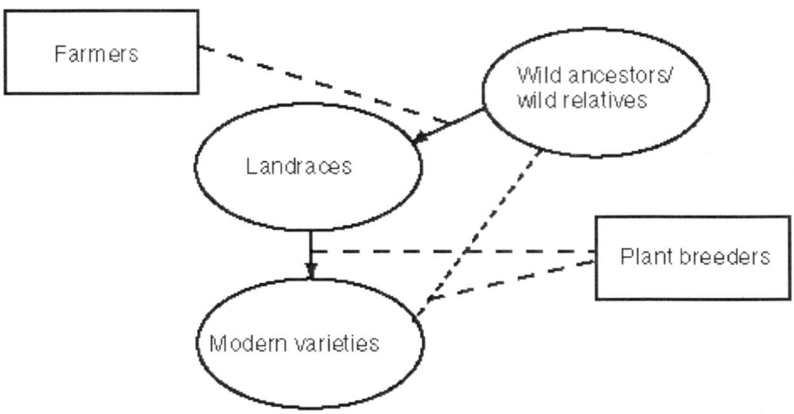

Figure 1. Farmers, plant breeders, and genetic resources.

Current Challenges

Changes in population, income, and other factors (such as urbanization) drive continuing increases in demand for agricultural commodities. Environmental conditions also change and pests and diseases evolve over time, so breeders continually need new and diverse germplasm from outside the utilized breeding stock, sometimes using wild relatives and landraces, to find specific traits to maintain or improve yields (Duvick, 1986). Maintaining resistance is a continual process, because new varieties are resistant to pests and diseases for an average of 5 years, while it generally takes 8 to 11 years to breed new varieties (USDA, 1990).

But private incentives to acquire and preserve genetic resources outside regular breeding stocks are limited, because genetic resources have strong "public goods" characteristics (Brown, 1987; Brown and Swierzbinski, 1985; Frisvold and Condon, 1994; Sedjo, 1992; Simpson and Sedjo, 1992; Reid, 1992; Swanson, 1996). For example, genetic resources are easily transported and replicated, and intellectual property protection has historically been relatively weak for biological innovations, making it difficult for an individual country, firm, or farmer to exclude others from their use. Furthermore, the usefulness of particular genetic resources is highly uncertain, and time horizons for improving genetic resources are long.

Despite the limits on private returns to their conservation and improvement, diverse crop genetic resources remain critical to agricultural production. Therefore, the public sector has played a pivotal role in their conservation. This raises three questions.

First, what are genetic resources worth? Most genetic resources are not market goods; that is, they are not sold as inputs into the breeding process and so lack simple indicators of their value. As such, policymakers find it difficult to compare investment in conservation with other uses for public funds. The international exchange of germplasm is also complicated, as countries may seek to maximize the returns from the set of resources that they hold.

Second, how diverse are genetic resources, not only in gene bank collections but also in the field? Diversity among genetic resources in the field can reduce the prospects for pest and disease epidemics. Farmers generally grow the most productive varieties (in terms of yield or quality), which may or may not be diverse. Society as a whole may prefer a higher level of diversity than farmers do. Incentives for land conservation, the breeding process, and access to modern varieties all can affect diversity in the field. Even the way in which diversity is defined can alter the assessment of benefits associated with different production and conservation decisions.

And finally, what can be done to ensure we have the crop genetic resources that we will need? The reliance of agriculture on these resources suggests the importance of continued preservation efforts. Policy instruments such as funding for *in situ* and *ex situ* conservation, intellectual property rights, and negotiated terms of transfer can be used to promote genetic resource conservation. While these policies can be implemented at the national level, genetic resources are found throughout the world. No nation has all the resources it wants or may need in the future. Thus, international coordination of genetic resource conservation is critical to meeting the longterm requirements of agricultural production.

Economic Values of Crop Genetic Resources

Attaching a value to genetic resources is a complex task. Describing the kinds of benefits associated with these resources is easier. The simplest benefit arises from the direct use of genetic resources: to produce food and fiber or to help create new varieties of crops and livestock. These direct uses are the focus of this chapter, although option value may also be an important motivation for their conservation.[1]

The ultimate direct-use benefits of crop genetic resources are measured in the increased output, higher quality, better resistance to pests, diseases, and other stress, and other characteristics found in improved crop varieties. These benefits derive not only from the genetic resources contained in precursor wild relatives, but also from the efforts of farmers who domesticated the crop and developed landraces through many years of selection; the work of collectors and gene banks that assembled and preserved genetic material in the form of landraces and wild relatives; and the work of plant breeders who have continued to develop and improve crop varieties.

Estimating the Benefits of Genetic Enhancement

Separating the contributions of breeders from the contributions of the germplasm with which they work is difficult. Thus, many studies have focused on the value of "genetic enhancement," or the value arising from both genetic material and its use by breeders. Most efforts to measure genetic enhancement have focused on specific crop breeding programs, using one of two related methods. The first measures benefits derived from a breeding program directly, and calculates rates of return to plant breeding efforts by comparing breeding program expenditures with their benefits. Many rate-of-return studies depend on the second method, some form of growth accounting. Growth accounting attempts to account for all factors affecting yields and then estimates the portion of the yield increase due to genetic enhancements.[2]

Rate-of-return studies sometimes base their estimates of the benefits from genetic enhancement on experimental estimates of yield gains. Plant breeders and other crop scientists may measure genetic gains in crop yield by conducting experiments that attempt to control for the effects of other inputs.[3] Although these studies focus specifically on genetic gains in yield, they do not always correctly value the economic benefits derived from the use of genetic resources for two reasons. First, yield trials that estimate genetic gains in yield are often conducted with input levels that farmers would not use or under environmental conditions that farmers would not face, in part because such experiments rely on control of other inputs for statistical validity. But plausible farmer responses in the face of changing technologies and market-environmental conditions suggest that yield gains in the field are likely to differ from experimental yield gains (Alston et al., 1995). Second, the resulting supply shifts for individual farmers would need to be aggregated to an industry supply shift in order to analyze economic costs and benefits to all producers and consumers.

Studies valuing the plant breeding component of genetic enhancement (see box, “Economic Studies of the Value of Genetic Enhancement”) consistently demonstrate its high utility in creating new varieties with higher yields and better resistance to disease. In most cases, too, the economic benefits of genetic enhancement far surpass the costs. These studies do differ in methodology, so the magnitude of estimated economic benefits is often not consistent across studies. Although Evenson and Gollin (1997) made some efforts to estimate the values of genetic resources directly, for the most part, valuation methodologies have not separated out the contribution made by plant breeding from the contributions of conserving genetic resources in farmers’ fields or in gene banks. Nor do most studies provide a detailed welfare analysis of costs and benefits across producers (including non-adopters) and consumers.

Economic Studies of the Value of Genetic Enhancement

Thirtle (1985) estimated the contributions of biological advances—which include both genetic enhancements and other land-saving technological change—in U.S. crop production using growth accounting (controlling for changes in other inputs such as fertilizers, machinery, and pesticides). Thirtle estimated that biological advances increased corn yields an average of 1.7 percent per year between 1939 and 1978; wheat 1.5 percent; soybeans 1.1 percent; and cotton 0.5 percent. Thirtle further concluded that biological improvements contributed to 50 percent of the yield growth of corn, 85 percent for soybeans, 75 percent for wheat, and 24 percent for cotton. In Thirtle's definition, however, biological improvements included both the use of improved varieties and other land-saving changes in agronomic practices.[1]

Byerlee and Traxler (1995) estimated a rate of return of 52 percent for joint international/national wheat breeding programs in developing countries. Pardey et al. (1996) also used rates of return, focusing on the spillover economic benefits of breeding research—i.e., benefits that accrue in regions or countries other than those originally targeted. They analyzed benefits in the United States (either to U.S. research programs or directly to U.S. farmers) from plant breeding research conducted in 2 of the 15 International Agricultural Research Centers (IARCs) that make up the Consultative Group on International Agricultural Research (CGIAR) system. Pardey et al. estimated returns on U.S. financial support to these two programs and found benefit-cost ratios for the United States of up to 48 to 1 for rice and 190 to 1 for wheat. Brennan et al. (1997) estimated that 64 percent of the genetic improvements to Australian rice came from international germplasm, and that the total Australian benefits of varietal yield improvement from 1962 to 1994 were $848 million (1994).

Evenson and Gollin (1997) estimated that without the International Network for the Genetic Evaluation of Rice, 20 improved varieties of rice would not have been released. The present value of that lost production over a 20-year period (the average length of time a rice variety is economically viable) was estimated to be $1.9 billion. Using a discount rate of 10 percent, the authors estimated that the present value of an added landrace (in a variety introduced by the program) was $50 million.

[1]Technically, Thirtle estimated the rate of land-saving biological-chemical technical change as an exponential time trend within a nested Cobb-Douglas/CES production function. In the same function, a different exponential time trend was used to estimate laborsaving mechanical technological change.

Frisvold et al. (2003) attempted to overcome some of the limitations of earlier studies by adding two features: a global welfare analysis, and a multi-market partial equilibrium model that could calculate the joint effects of genetic improvements in five major crops in the United States between 1975 and 1992.[4] They first estimated the size and distribution of the gross annual benefits of a single-year increase (figure 2, first panel) in the U.S. yields of corn, soybeans, wheat, cotton, and sorghum. About half of the increase in yields can be attributed to improved seed varieties (Fuglie et al., 1996).[5] Accordingly, to simulate the effects of genetic improvement only, the authors increased the supply of crops by half of the average annual yield growth, implicitly assuming no changes in other inputs and no interactions between genetic improvements and other inputs.

Frisvold et al. estimated that the overall economic welfare of U.S. crop producers across the five commodities increased by more than $160 million and that consumer welfare increased by more than $220 million (1989 constant dollars) due to U.S. genetic improvements. Total U.S. economic welfare increased over $350 million. Producers in the rest of the world suffered losses, while consumers in the rest of the world gained from lower world food prices global welfare increased by $590 million, with the United States capturing 60 percent of the total gain, other developed countries 25 percent, and developing and transitional economies 16 percent.

In fact, yield increases from genetic improvements are not limited to a single year, so Frisvold et al. also calculated the present value of a permanent increase in yields from genetic improvements (figure 2, second panel).[6] The U.S. benefits of permanent U.S. yield increases range from just under $5 billion (1997 dollars) to over $9 billion. Global benefits range from $8 billion to $15 billion and benefits to developing and transitional economies range from $1 billion to $2.5 billion. (Consumer benefits in developing and transitional economies range from $6 billion to over $11 billion.)

estimates are conservative for two reasons. First, growth in income and population over time would make the total benefits of yield increases even larger as demand grows. And second, "plant breeding and genetic improvements that have not merely generated one-time permanent increases in yields, but rather an annual stream of permanent yield improvements. Every year there is a new incremental permanent increase in yields. The problem is equivalent to receiving a new annuity of varying value every year" (Frisvold et al., 2003) (figure 2, third panel).[7]

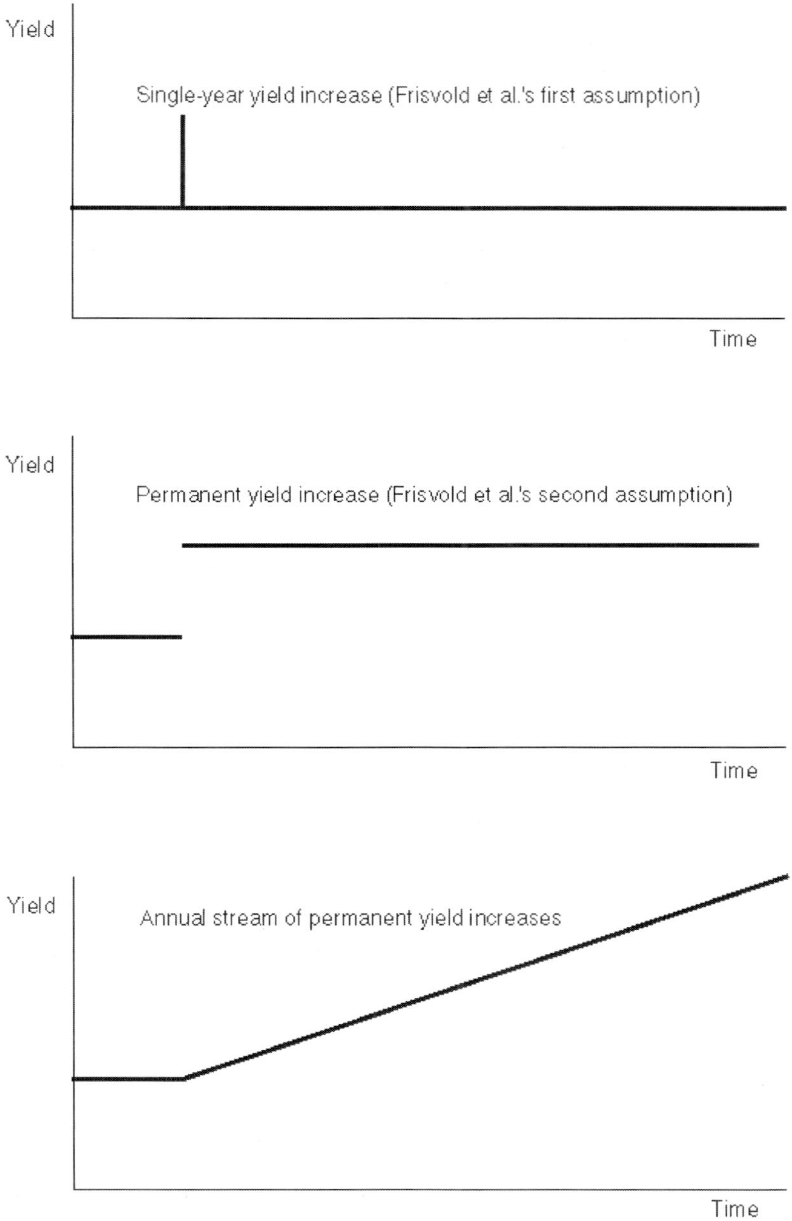

Figure 2. Alternative assumptions about benefits from genetic enhancement.

These These results suggest that investment in genetic enhancement has generated large returns. The United States was the major beneficiary of genetic enhancement in U.S. crops, although the genetic resources used in these improvements might have multiple sources. (Note that Frisvold et al's analysis did not include the U.S. research costs necessary to achieve these yield gains.) Nonetheless, developing and transitional economies also benefited from U.S. yield gains, and it is likely that poor consumers in these countries (including many small farmers) are among the major beneficiaries.

Searching for Valuable Genetic Resources

Genetic enhancement depends on the availability of diverse genetic resources for use by plant breeders. In addition to evaluating genetic enhancement, economists have also attempted to evaluate the search for agricultural genetic resources *in situ* (in their natural habitat), the storage and characterization of these resources *ex situ* (e.g. in germplasm collections), and the search for particular traits within *ex situ* collections. Compared with estimates of returns to genetic enhancement, estimates of search costs and returns often are more complex conceptually and more demanding of scarce data (see box, "Economic Models of Searching for Genetic Resources").

Most models of the economics of searching for genetic resources held *in situ* or *ex situ* have been difficult to apply empirically due to data limitations. Several different types of empirical studies have, however, provided useful information about the economics of conservation. First, Evenson and Gollin (1997) directly estimated likely benefits of additional accessions to the rice collection maintained by the International Network for the Genetic Evaluation of Rice. They estimated that the present value of 1,000 additional accessions (discounted at 10 percent over a 20-year period) was $325 million.

Second, Pardey et al. (2001; 2004) estimated the marginal costs of adding accessions to the *ex situ* gene bank for wheat and maize (corn) at CIMMYT, the International Maize and Wheat Improvement Center, and estimated the cost of holding an additional accession in perpetuity. Though Pardey et al. did not estimate the expected values of benefits for additional accessions (and suggested it might not even be feasible), they argued that the cost of additional wheat accessions was so low that expected benefits would probably always outweigh this cost. They argued that some accessions to the maize gene bank—e.g., landraces and wild relatives—might be more likely to have an expected positive return than others, like recently created breeding lines. This is because useful genetic material contained in breeding lines might well be conserved elsewhere—for example, by maize breeding programs—but useful genetic material in landraces and wild relatives would probably be conserved only in the gene bank.[8]

Third, surveys of plant breeders and other users of gene banks have consistently showed that they find gene bank materials useful. For example, the U.S. National Plant Germplasm System (NPGS) is one of the largest national gene banks in the world; it distributes, for free, more germplasm samples internationally than any other supplier, including the international research centers of the CGIAR. Smale and Day-Rubenstein (2002) found that international users of the NPGS requested materials for a variety of uses, including basic research and breeding, and a majority expected that their use of NPGS materials would stay the same or increase in the future. Of NPGS samples distributed from 1995 through 1999, 11 percent had been used in breeding programs, 18 percent were found useful in other ways, and 43 percent were still being evaluated. Twenty-eight percent of the samples were not considered useful. Rejesus et al. (1996) found that wheat breeders around the world used released cultivars, advanced materials, and germplasm from international nurseries much more frequently than wild relatives and landraces. Wild relatives and landraces were used particularly in search of specific traits, such as disease resistance, drought resistance, and quality.

Economic Models of Searching for Genetic Resources

Simpson, Sedjo, and Reid (1996) applied a theoretical model, originally used in labor economics, to biodiversity conservation in the context of a search for species of interest to pharmaceutical research. Modifying this model, Simpson and Sedjo (1998) argued that the value to society of biodiversity prospecting (searching for genetic resources currently held *in situ*) for use in crop improvement programs was likely to be low.

Cooper (1998) approached the question as one of investment in "converting" *in situ* genetic resources into *ex situ* resources under (1) uncertainty concerning the measurement and value of *in situ* genetic resources, and (2) irreversibility since *in situ* resources, once lost, cannot be replaced. Cooper's simulations demonstrated that estimates of mean benefits might not be particularly useful, as the range of potential benefits could be quite large.

Evenson and Lemarié (1998) applied a search model to a two-stage process—first, collecting genetic resources *in situ* and placing them *ex situ,* and second, searching the *ex situ* collection for traits of interest. They showed that the optimal size of a collection depends on the number of traits being sought, and on the distribution of genetic resources across geographic regions.

Gollin, Smale, and Skovmand (2000) developed a theoretical model that characterizes the search for resistance to pests and diseases in *ex situ* collections of wheat genetic resources, and then analyzed data on frequency distributions, disease losses, and search costs. They concluded that "the optimal size of search for traits is highly sensitive to the economic magnitude of the problem, the research time lag, and the probability distribution of the trait." Furthermore, even though subcollections of landraces or wild relatives might be used only on rare occasions, high benefits might result on those occasions. The fact that "gene banks and some categories of accessions"—i.e., certain types of genetic materials held by a gene bank—"are infrequently demanded by crop breeders does not in itself imply that marginal accessions have low value."

Drawing on these earlier studies, Rausser and Small (2000) argued that scientific models that "channel research effort towards leads for which the expected productivity of discoveries is highest" significantly reduce search costs from earlier "brute force" models that assume no prior information can be brought to the search. In contrast to the results of Simpson et al., Rausser and Small's simulations suggest that market-based conservation of genetic resources might be possible in some cases because prior information reduces private search costs so they are lower than expected private benefits from searching.

One final consideration refers not to the economics of plant genetic resource conservation *per se,* but to a related scientific development bearing on economic decisionmaking. This is the potential of modern molecular biology, including genomics, to reduce the search costs for useful traits in conserved material. (Genomics refers to investigations into the structure and function of very large numbers of genes undertaken simultaneously.) At this point, however, it is relatively easy to generate mountains of raw genetic sequence data but difficult to transform these data into useful information (Attwood, 2000). Thus, conserved genetic resources may increase in value as genomics and other molecular techniques lower search costs and the costs of capitalizing on search results, but it is difficult to predict the pace at which this will take place.

The literature on searching for valuable genetic resources is less conclusive than the literature on evaluating the benefits of genetic enhancement. The majority of studies agree that economic benefits from searching for genetic resources either *in situ* or *ex situ* are positive compared with costs. This can be true even if successful searches are a small fraction of the total searches conducted. However, studies also conclude that it is quite difficult to value searches for genetic resources, and that the range of potential values may be large. The key variable is information. Application of prior information about the probability distribution of a desired trait or set of traits and where searches are likely to have the highest payoffs can significantly increase the economic value of a search for genetic resources. This prior information might be embodied in knowledgeable individuals, scientific publications, characterization of gene bank holdings, or the findings of molecular biology.

Taken together, economic analysis of genetic enhancement and the search for genetic resources indicate that returns to the discovery and use of crop genetic resources exceed the costs. Many scientists, however, have raised concerns about the continued availability of sufficient genetic resources for future plant breeding efforts. Furthermore, both the scientific and economic literatures agree that the measurement of genetic diversity is complex.

Factors Influencing Trends in Crop Genetic Diversity

Diverse genetic resources have been a source of large gains in agricultural productivity and, as a result, producer and consumer well-being. Such gains might provide incentives for conservation and efficient use of valuable resources, but these incentives are often muted in the case of genetic resources because returns to their identification and use are not always easily captured by individual farmers, firms, or countries. In fact, the loss of genetic diversity in a species, also called genetic erosion, has been reported in many commercially important crops (National Research Council, 1972; National Research Council 1993; Porceddu et al., 1988).

Genetic diversity is a particular concern because greater genetic uniformity in crops can increase vulnerability to pests and diseases (National

Research Council, 1993). Genetic uniformity does not, in and of itself, mean that a particular variety is more vulnerable to pests and diseases or abiotic stresses. In fact, modern varieties often are bred for superior resistance, hence their popularity. Nonetheless, as pests and diseases evolve to overcome host plant resistance, genetic uniformity increases the likelihood that such a mutation eventually will prove harmful to a crop. The evolved pest or disease has a greater crop base that it can successfully attack, which could increase its severity. Instead of a particular disease harming only a small percentage of varieties on limited land, the disease now could affect a greater proportion of a crop's production. For example, genetic uniformity contributed to the spread of the Southern Corn Leaf Blight, which led to a 15-percent reduction in the U.S. corn crop in 1970.

Here, we identify three factors that might contribute to loss of genetic diversity—habitat loss, conversion from landraces to scientifically bred varieties, and genetic uniformity in scientifically bred varieties—and assess how much each factor is operative today. Considerable debate surrounds both the historic and current loss of genetic diversity, due in part to difficulties in defining an appropriate concept of genetic diversity and obtaining accurate measurements. (Formal measures of genetic diversity, as applied both by scientists and by economists, are discussed in the Appendix.) Formal measures of genetic diversity tend to be both wide ranging and data-intensive, and, in most cases, they are not available for long periods (see box, "Measures of Crop Diversity"). As a result, the discussion of trends in genetic diversity is indicative, not precise.

Most of the formal definitions of genetic diversity are applied either at the cross-species level or within a particular species. Within a crop species, these definitions may be related to the number of varieties, the distribution of varieties within a given area, and/or the genetic difference between varieties within a given area or period of time. In the context of crop genetic resources, for example, habitat loss is likely to affect diversity primarily at the cross-species level, where the relevant species are those closely related to the crop of interest. Conversion from landraces to scientifically bred varieties and genetic uniformity in scientifically bred varieties, on the other hand, may affect one or more of these types of indicators *within* a particular crop species.

Habitat Loss

One factor contributing to a decline in crop genetic diversity has been the loss of wild relatives of cultivated crops (National Research Council, 1993). The loss of wild relatives occurs mainly through habitat conversion for agricultural use. When forest and other wild lands are cleared, plant, animal, and microorganism populations generally fall, reducing the level of genetic diversity. Habitat loss is particularly problematic in developing countries, which often face greater pressures for wild land conversion than do developed countries (Houghton, 1994). Population growth and extensive farming techniques are often cited as factors fostering high rates of land conversion to agriculture. Other influences on land conversion are thought to include poverty, international trade, land degradation, and government policies, particularly where land tenure policies are not clearly defined or enforced (Day-Rubenstein et al., 2000).

Because the full economic values of wild relatives can rarely be captured by landowners, the use of land to preserve habitats for wild relatives remains undervalued compared with alternative uses such as clearing for agricultural or urban use. Thus, habitat conversion occurs in part because the private returns to genetic and other biological diversity are lower than the social returns (Hanemann, 1988). Private returns are important because resources are generally held (whether formally or informally) at the individual or local level. Therefore, many decisions that affect conservation of biodiversity, such as land clearing, are made at these levels. By contrast, many of the benefits of biodiversity conservation accrue at the national or global level. These differing returns contribute to biological resource depletion because conservation of habitat competes with alternative uses of land. Since keeping land in its natural state reduces or eliminates the land's earning capacity for its holders, returns to agricultural production form one opportunity cost of wild land preservation. Also, temporal issues come into play: individuals may place a greater value on current consumption, when weighing the tradeoff between present and future use of resources, than does society as a whole. Together, these factors generate private or individual decisions that differ from those that are socially or globally optimal.

MEASURES OF GENETIC DIVERSITY

Measures of genetic diversity are very numerous, although there are strong similarities and relationships among many of these measures. At a general level, most involve measures of the *number* of species, the *distribution* of species, and/or the *difference between* species within a given area or period of time. More narrowly, similar concepts might be applied within a crop species, with varieties rather than species becoming the relevant unit of observation.

One reason for the wide variety of measures of genetic diversity is that different people have different reasons for studying or using it. Evolutionary biologists might want to study the process of speciation or the formation of new species, or measure the evolutionary distance between species. Ecologists may be interested in the number and distribution of species within a given habitat. Plant breeders usually focus more closely on diversity within a crop species of interest, although they may also wish to tap diversity within the secondary and tertiary gene pools for that species. (The *secondary gene pool* consists of all biological species that can be crossed with the cultivated species, although these crosses are usually sterile. The *tertiary gene pool* consists of those species that can be crossed with the cultivated species only with difficulty, such as with genetic engineering).

Farmers, particularly those cultivating landraces in noncommercialized agriculture, may be interested in morphological diversity—i.e., diversity in certain physical traits. Because traits are influenced by environmental factors, and because, in many cases, many interacting genes contribute to trait expression, morphological diversity may not be considered to be a "true" measure of genetic diversity. Nonetheless, farmers may make their planting decisions based on such morphological diversity, so it is a potential influence on underlying genetic diversity. Policymakers may focus on preserving genetic diversity as a means to continue crop improvement and guard against the risks of pest or disease epidemics. Economists may wish to study the ways in which the variables important to farmers or policymakers interact with the variables important to plant breeders or ecologists. But no single measure fulfills all desired criteria (Meng et al., 1998).

Also, because certain genetic materials are easy to transport and replicate once collected, it is difficult for countries to capture more than a fraction of the value that flows from their genetic resources. Moreover, markets do not exist for most of the other environmental services provided by biological resources, such as carbon sequestration. Consequently, keeping land in less intensive uses favorable to the *in situ* preservation of genetic resources is often less profitable than more intensive agricultural production to individual countries as well as to individual landowners.

Although many habitat reserves have been established worldwide, wild relatives of agricultural species tend to be included only by accident (FAO, 1996b). Habitat preserves often focus on areas rich in species diversity—usually wildlife species or all plant species—and not on crop species alone. These areas are not necessarily those with the greatest crop genetic diversity.

Much empirical work has focused on the loss of tropical forests, but continued agricultural expansion onto other land is also expected (Day-Rubenstein et al., 2000), although at rates lower than previously projected (Bruinsma, 2003). Compared with the developing world, the developed world has lower rates of agricultural land expansion. For example, the amount of U.S. land used for agricultural production has remained stable since 1945 (ERS, 2002). This does not mean that the same land has been in production. Urban land expansion has displaced some agricultural lands, which have displaced some wild lands. Still, expansion of the agricultural production area has not been a significant factor in U.S. biodiversity loss in recent years.

Displacement of Landraces by Scientifically Bred Varieties

Crop genetic diversity also declines as landraces are displaced by scientifically developed modern varieties (National Research Council, 1972; Proceddu et al., 1988; Chang, 1994; Kloppenburg, 1988). The ongoing selection process is thought to have narrowed the genetic base of varieties used in agricultural production (Brush, 1992; FAO, 1996b; GAO, 1997; Goodman and Castillo-Gonzalez, 1991). In particular, the spread of high- yielding "Green Revolution" varieties and associated changes in crop management practices beginning in the 1960s is thought to exemplify this transition from landraces to modern varieties (Frankel, 1970; Tilman, 1998). Far less area is planted to landraces worldwide than a century ago. But in many cases, the transition to modern varieties predates the Green Revolution. Improved crop varieties, such as hybrid corn or semi-dwarf wheat or rice, often replaced other varieties that were already the products of scientific crop improvement (see Smale, 1997, for an example). In the broadest sense, alteration and narrowing of crop genetic diversity began with the first domestication of wild plants. For example, the corn plant has been completely dependent on humans for reproduction for thousands of years, because farmer selection has resulted in kernels that can no longer disperse without human intervention.

Farmer choice is a key driving factor behind the replacement of landraces with scientifically bred varieties. When choosing varieties, farmers consider yield potential as well as other production and consumption attributes. Sometimes landraces offer superior yields or resistance to biotic and abiotic stresses, but often they do not. Landraces often provide consumption characteristics traditionally preferred to those of modern varieties (such as maize better suited for tortillas), but even this advantage is not absolute. While maintenance of a diverse set of landrace varieties may prove valuable to current or future plant breeding, individual farmers do not directly capture these benefits, so they have little incentive to account for them when selecting seed for planting. Landraces become extinct through disuse if farmers stop planting and maintaining them, unless stored *ex situ*. Even if many landraces are stored in gene banks, genetic diversity might be lower than if these landraces were planted by farmers, because in the gene bank they are not subject to ongoing evolutionary pressure.

The rate of landrace replacement by scientifically bred varieties differs by crop, world region, and environment. In most industrialized nations, commercialized crops—i.e., crops grown solely for the market, not home consumption—consist almost completely of scientifically bred varieties, although isolated use of landraces may occur.[9] In developing countries, genetic resource specialists often have information about the location of crop landraces and the rate at which they are being replaced by scientifically bred varieties, but published information that is accurate and aggregated is difficult to find.

Some information is available, however, for use of landraces of the three major world cereals, rice, wheat, and corn (maize). In the 1990s, approximately 15 percent of the global area devoted to rice was planted to landraces. Rice landraces are concentrated in southeast Asia, with some also found in the Indian subcontinent (Cabanilla et al., 1999). Use of rice landraces varies by environment and is much lower in the irrigated lowlands than in the more difficult rain-fed lowland and flood-prone and upland environments.

About 10 percent of the developing world's wheat area was planted to land- races in the 1990s. Wheat landraces were concentrated in West Asia and North Africa, with some also found in Ethiopia, China, the Indian subcontinent, and small areas in Latin America. The proportion of wheat area planted to landraces also varied by wheat type and environment. For example,

23 percent of the area planted to durum wheat and 12 percent of the area planted to winter bread wheat was sown to landraces, while only 3 percent of the spring bread wheat area in developing countries was still planted to lan-draces (Heisey et al., 2002).

Unlike wheat and rice, which self-pollinate, corn cross-pollinates, which means that one plant is often fertilized by another. Because of this feature, corn populations are inherently less stable genetically. Therefore, corn landraces may be very diverse genetically. Furthermore, if farmers continue to replant seed (even from hybrids or other scientifically improved corn varieties) rather than buying new seed, the resulting progeny may also be quite genetically diverse. As a result, it is more difficult to define and measure what constitutes a landrace and what is "improved germplasm" for corn than it is for rice or wheat (Morris et al., 1999). That said, it is clear that a far higher percentage of the developing world's corn area (just under 40 percent) is planted to landraces than is the case for either wheat or rice. If developing countries that produce primarily temperate corn or countries that market "commercialized" corn[10] are excluded, nearly 60 percent of the developing world's corn area is planted to landraces (Morris, 2002). As with the other cereals, corn's wild relatives tend to concentrate in their zone of origin (in the case of corn, in Mexico and Central America), and landraces are most diverse in this zone. Nonetheless, corn landraces are found in many parts of the developing world.

Genetic Uniformity in Scientifically Bred Varieties

In situations where most or all landraces have been replaced by scientifically bred varieties, crop genetic diversity may also decline with (1) reductions in total numbers of varieties, (2) concentration of area planted in a few favored varieties, or (3) reductions in the genetic distance between these varieties. The National Research Council (1993) concluded that the genetic vulnerability of U.S. wheat and corn has become less of a problem since 1970, in part because of efforts to breed in greater diversity. However, the Council also determined that genetic uniformity of rice, beans, and many minor crops is still a concern.

Information for other countries is not readily available. Relatively little attention has been paid to genetic uniformity of scientifically bred varieties in developing countries, perhaps because there more focus has been placed on habitat conversion and displacement of landraces. One major study, however, analyzed trends in modern spring bread wheats planted in the developing world, both in the genetic diversity of varieties released and varieties planted in farmers' fields (Smale et al., 2001). This study was representative of over 50 million hectares of wheat planted in the developing world. Both pedigree analysis and molecular analysis suggested that the genetic diversity of these modern wheat varieties had increased, not decreased, over the past 30 years. Trends in genetic diversity for other crops in developing countries, however, as well as for crops in industrialized nations outside the United States, would likely vary by crop and region. Whatever the trends in genetic diversity, the genetic uniformity of many crops has raised concerns that crop yields and production will become more variable from season to season (Swanson, 1996). As with other drivers of genetic erosion, individual farmers have limited incentives to consider the wider potential consequences of genetic uniformity, and, when choosing which varieties to plant, may perceive the benefits of uniform varieties to be greater. Farmers may be willing to accept the risk of greater variability if they expect to receive higher average yields.

Thus far, despite concerns about genetic uniformity, yields for many major crops have been relatively stable. An important reason may be that temporal diversity has replaced spatial diversity (Duvick, 1984). Although there may be greater spatial uniformity of crops planted at any given time today (compared with 100 years ago), modern plant breeding provides a steady release of new varieties with new traits for pest or disease resistance over time.

The ability of plant breeders to keep ahead of evolving pests and diseases through temporal diversity depends directly on the quality and accessibility of germplasm collections in public gene banks and in private breeders' collections. Because many of the benefits of raw germplasm cannot be appropriated, private breeders rely on the public sector to collect, characterize and perform pre-breeding enhancement of genetic materials to make them accessible for private use (Duvick, 1991).

CONSERVATION OF PLANT GENETIC RESOURCES

In this section we examine two basic strategies for conserving genetic resources, three principal tools policymakers can use to support these strategies, and several multilateral agreements by which countries currently seek to coordinate international use of these tools. Decisions about these alternatives may affect U.S. access to genetic resources that are currently held outside the United States (and vice versa).

Table 1. Advantages and disadvantages of *ex situ* versus *in situ* conservation

Ex situ conservation		*In situ* conservation	
Advantages	**Disadvantages**	**Advantages**	**Disadvantages**
Costs generally centralized	Certain types of germplasm not readily conserved	Genetic resour-cees used to pro-duce valuable product	Costs borne by farmers (for landraces)
Can preserve large amounts of diverse germplasm	Regeneration can be costly, time-consuming	Evolutionary processes continue	May reduce on- farm productivity
Germplasm can be readily accessed by more breeders	Potential for genetic "drift" can reduce integrity of collection	May better meet the needs of certain farmers	Requires land
High-security storage imper-vious to most natural disasters.	In practice, many collections lack the resources nee-ded to organize, document, and maintain their samples.	More efficient for some germ-plasm, e.g., animals, or crops that reproduce vegetatively.	Farmer selections may not preserve targeted diversity
		Existing wild relatives can be preserved with-out collection	Loss of wild relatives when land use changes

Note: The pointer locations indicate general regions where crops are believed to have first been domesticated. In some cases, the center of origin is uncertain. Other geographic regions also harbor important genetic diversity for these crops.

Source: This map was developed by the General Accounting Office using data provided by the National Plant Germplasm System's Plant Exchange Office.

Figure 3. Centers of origin of selected crops.

Basic Conservation Strategies

At the most basic level, genetic resources can be conserved either *in situ* (in their natural setting) or *ex situ* (outside their natural setting). *In situ* is the dominant method of conserving natural ecosystems. Crop genetic resources are commonly held *ex situ*, but they can also be held *in situ*—as wild relatives of cultivated varieties on wild land and as cultivated varieties in farmers' fields. Among the decisions policymakers face is the appropriate balance between *in situ* and *ex situ* conservation efforts. Each has its own benefits and drawbacks; the two are perhaps better viewed as complementary rather than as substitutes (table 1).

In Situ Conservation

Species preserved *in situ* remain in their natural habitat. Most of the world's genetic diversity is found *in situ*. For agriculturally important species, the greatest diversity in landraces and in wild relatives is typically found near where they were first domesticated. Early in the twentieth century, Russian botanist N. I. Vavilov defined "centers of origin" for most crops. These included Mexico and Central America (for corn, or maize as it is known in the rest of the world, and upland cotton); China (for soybeans); and West Asia (for wheat and alfalfa).

Since Vavilov's time, ideas about centers of origin have been refined. Some crops, such as sorghum, sugarcane, and peanuts, were probably domesticated over very broad areas rather than in a well defined center (Harlan, 1971, 1992). Furthermore, useful landraces of some crops have been found in parts of the world other than those in which they were originally domesticated. For example, wheat landraces found in the pedigrees of many modern wheat varieties have come from every continent except Antarctica (Smale and McBride, 1996).[11] Still, *in situ* preservation efforts, as well as germplasm collection activities for *ex situ* conservation, are often focused most closely in and around centers of origin (figure 3).

Because *in situ* conservation of agricultural genetic resources is carried out within the ecosystems of farmers' fields or wild lands, species continue to evolve with changing environmental conditions. *In situ* conservation thus can provide valuable knowledge about a species' development and evolutionary processes, as well as how species interact. By allowing genetic resources to act as part of larger ecosystems, *in situ* conservation may also provide indirect ecological benefits, such as hosting diverse pollinators. However, since restrictions on land use may be necessary, *in situ* conservation can be costly. To conserve agricultural genetic diversity *in situ,* for example, a farmer may have to forgo the opportunity to grow a higher yielding (and more profitable) variety. Or, in the case of wild *in situ* resources, the land may need to be set aside from agricultural production or other production-related uses completely. This suggests one important constraint on *in situ* conservation that has been addressed in our discussion of habitat loss—the divergence between the social and private returns to conserving genetic diversity.

Ex Situ Conservation

The *ex situ* method removes genetic material from its environment for longterm conservation (table 1). Botanical gardens and gene banks are examples of *ex situ* conservation strategies. Certain methods of *ex situ* conservation can be used to store large amounts of genetic material at relatively low cost, certainly in terms of land needed, compared with *in situ* strategies. The world's gene banks presently hold more than four million accessions, or specific samples of crop varieties. It is estimated that samples of many of the world's cereal landraces are now held in gene banks (Plucknett et al., 1987). Although very few important crop species originated in what is now the United States, the U.S. national gene bank system (the National Plant Germplasm System, or NPGS) is today one of the largest *ex situ* collections in the world. *Ex situ* conservation also is appealing because it allows plant breeders easier access to genetic resources than is provided by *in situ* conservation.

However, crop genetic resources first must be collected, and samples of only a small fraction of the world's plant genetic resources have been collected thus far. Stored plant materials must be kept under controlled conditions, and periodically regenerated (planted and grown) in order to maintain seed viability. Not all kinds of plant genetic resources are easily conserved *ex situ*. Some lose their varietal identity when stored as seed. These plants may need to be kept as living plants, a more costly process that requires additional land and labor. And gene banks in politically unstable areas may be in danger of losing valuable genetic material. Even in stable locations, the resources necessary to maintain or improve plant gene banks are not always forthcoming because of competing demands for public resources (GAO, 1997).

Policy Tools to Promote Genetic Resource Conservation

Three major types of policy tools are available to support conservation of genetic resources: (1) public investment in *in situ* and *ex situ* conservation; (2)

stronger intellectual property rights over genetic inventions, particularly in developing countries; and (3) material transfer agreements.

Public Funding of Ex Situ and In Situ Conservation

Funding conservation is the most direct method of preserving crop genetic resources. Past efforts have convinced plant breeders that the current germplasm stock, if properly maintained, is adequate to maintain steady yield growth over the next 20 to 50 years (Shands, 1994; Sperling, 1994; Siebeck, 1994). There is growing concern, however, that this may not be sustainable in the long term at current funding levels (Keystone Center, 1991; NRC, 1993; OTA, 1987; FAO, 1996b). Studies of gene banks worldwide (FAO, 1996a), the U.S. National Plant Germplasm System (GAO, 1997), and the Vavilov Institute collection in the former Soviet Union (Zohrabian, 1995) conclude that most gene banks lack sufficient funds, facilities, and staff to maintain their germplasm collections.[12] Funding problems arise, in part, because individual nations do not capture the full benefits of investments in genetic resource conservation. While multilateral funding of international crop research facilities has been used to alleviate this problem, free rider problems suggest that funding for international facilities will remain less than optimal.

The UN Food and Agriculture Organization (FAO) reported on the most pervasive problems facing gene banks worldwide (FAO, 1996a). First, since 1970, more emphasis has been placed on collecting materials, than maintaining accessions, and most gene banks lack adequate long term storage facilities. Even accessions in suitable long term storage cannot be maintained indefinitely; collected material must be grown out or "regenerated" periodically. Many gene banks lack the funds, facilities, or staff to carry out needed regenerations. Second, while gene bank coverage of elite and landrace varieties of major cereal crops is believed to be fairly complete, coverage of many "minor" crops (such as root crops, fruits, and vegetables) and wild relatives remains spotty. Third, only a small fraction of accessions has been characterized. This lack of information about what actually resides in these collections constrains breeders from using new genetic materials (NRC, 1993) and makes it difficult to identify gaps in collections. Fourth, many countries have reported that funding has been unstable and uncertain year to year, hampering investment and planning decisions. The FAO (1996c) concluded that "without prompt and significant intervention, much of the stored genetic diversity of food and agricultural crops in the world-as well as the large public investment made in assembling the collections-will be lost forever."[13]

The same public goods problem that inhibits optimal international investment in *ex situ* conservation of genetic resources-the inability of conserving nations to capture all the benefits from that conservation-also hinders optimal investment in *in situ* conservation. Moreover, *in situ* conservation is subject to several additional constraints. First, uncertainty surrounding the likely magnitudes of the benefits of in situ conservation is probably larger than it is for *ex situ* conservation. Second, the number of economic agents and levels involved in any *in situ* conservation effort (including landowners and/or individuals with rights to use the land) is likely to be considerably larger than for *ex situ* programs, making coordination of *in situ* programs more difficult.

In situ conservation of wild relatives and landraces require different strategies. Establishing habitat reserves could protect wild relatives. Turkey, for example, has received multilateral funding for an *in situ* pilot project to conserve wild relatives of wheat and barley (FAO, 1996b). For landraces, if farmers have private incentives to maintain local varieties, policy interventions for *in situ* conservation may be unnecessary. In areas where displacement of local varieties is more likely, access to modern varieties need not be completely prohibited. A less costly alternative might be to establish some type of conservation easement, paying local farmers the difference between returns to modern and local varieties if they grow a diverse set of varieties on part of their plots (Christensen, 1987). Yet another approach could be to purchase limited amounts of landrace seed from producers in regions with diversity

Most experts agree that *in situ* and *ex situ* conservation strategies are complementary, however the best allocation of resources is subject to debate. Plant breeders are concerned that increased investment in *in situ* conservation will compromise gene bank maintenance. Lack of data on the relative costs and benefits of *in situ* and *ex situ* conservation increases the difficulty of allocating funds across activities. Moreover, donor institutions, particularly at the national level, face competing needs, some of which offer more direct and immediate benefits.

Intellectual Property Rights

Adoption of stronger intellectual property rights (IPR) regimes has been one of the most commonly proposed methods to enhance genetic resource conservation internationally. Proponents argue that stronger IPR will allow the holders of genetic resources to reap the rewards from commercializing these resources and thus align private incentives more closely with public incentives for genetic resource conservation.

Historically, the set of IPR used for genetic resources internationally focused on the products of formal plant breeding programs rather than wild relatives and landraces. Even while varieties developed by breeders were protected by formal "plant breeders' rights", wild relatives and landraces continued to be considered a public good. For decades, many plant breeders have freely exchanged "raw" germplasm (Kronstad, 1996; Heisey et al., 2001).[14] National plant breeding programs and international agricultural research centers freely provide such unshielded genetic materials not only to other public breeding institutions but also to private breeders (many of them in developed countries) who may then use those materials to develop new commercial crop varieties for sale (Day, 1997).[15]

This asymmetry has proven controversial. Many developing countries and nongovernmental organizations (NGOs) make the case for "farmers' rights," arguing that farmers in developing countries have selected and saved landraces for thousands of years, making an essential contribution to plant breeding and crop variety development (Mooney, 1979, 1983; Brush, 1992). It is unfair, they argue, that private breeders have free use of wild relatives and landraces but require payment for elite varieties based, in part, on germplasm that originated in developing countries. Others counter that the exchange of genetic material for plant breeding has been beneficial to developed and developing countries alike, although they disagree about whether foregone earnings from sales of raw genetic material by lower-income countries are compensated for by other benefits, such as unrestricted access to public germplasm and lower food prices for consumers (Shands and Stoner, 1997; Fowler, 1991).

Proponents of stronger IPR regimes argue that, generally speaking, they encourage commercialization of genetic resources, thus enhancing the incentives for conservation, both *in situ* and *ex situ*. They also maintain that greater IPR stimulate private sector research, relieve public budgetary constraints, and increase national incentives for germplasm conservation (Barton and Siebeck, 1991). Critics counter that stronger IPR would do little to increase innovation or maintain crop genetic diversity, arguing that private incentives favor specialization and product uniformity rather than diversity in the production of new seed varieties (Mooney, 1979, 1983; Acharya, 1991; Reid, 1992; Brush, 1994).

These arguments raise two empirical questions. First, what impact would stronger IPR protection have on germplasm use and exchange? A survey of 84 private plant breeding firms by Pray et al. (1993) assessed the impacts of a 1985 decision by the U.S. Supreme Court that strengthened genetic resource IPR (for modern varieties) by allowing plant breeders to acquire utility patents for new varieties.[16] More than a third of the firms felt that utility patents limited germplasm exchange both between private firms and between the public sector and private firms. Six of 84 firms reported that they had increased their research expenditures because of the availability of utility patent protection. Most reported that utility patents increased profitability. Rejesus et al. (1996) surveyed wheat breeders internationally, and reported that respondents believed that stronger international IPR for plant varieties would reduce germplasm exchange between developed and developing countries, reduce exchange between developing countries and reduce the use of foreign landraces. Pray (1990) noted that stronger IPR in developing countries would entail significant enforcement costs and other transaction costs. The effect of IPR targeted toward land races and wild relatives remains unknown.

Second, what are the implications of stronger IPR and increased private R&D for the diversity of new varieties developed? Some evidence suggests that the diversity of major crops has not declined in the United States as increasingly strong IPR protections have been enacted over the last 30 years, and diversity may have actually increased for some crops (Duvick, 1984; NRC, 1993; Smale and McBride, 1996; Falck Zepeda and Traxler, 1997; Pray and Knudson, 1994; Knudson, 1998). But the role of IPR is confounded by other efforts to increase crop genetic diversity (see NRC, 1993, pp. 67-81, for discussion on the impacts of its 1972 report "Genetic Vulnerability of Major Crops").

Material Transfer Agreements

Material transfer agreements (MTAs) are legal instruments initially used as a means for transferring biological materials between entities, including public institutions, private companies, and countries. Initially, used for research only, MTAs may be bilateral agreements or may follow a standard template (such agreements are often used by public entities). The provider retains commercial rights to the material. MTAs have become a common instrument to outline the terms for sharing genetic resources and, sometimes, the gains from new product development. MTAs may include provisions for intellectual property rights, such as what, if any, IPR may be sought for the transferred material or inventions based on that material. However, not all MTAs address IPR and even if they do, IPR usually are just one element of the agreement.

Interest in using MTAs as an incentive to preserve germplasm stems from the idea that benefit sharing can reward suppliers of genetic resources (Barton and Christensen, 1988; Blum, 1993; Christensen, 1987; Simpson and Sedjo, 1992; U.S. Department of State, 1994; WRI, 1993). The benefits to be shared may include funds, materials, training, technology, or intellectual property rights (through provisions concerning their allocation).

The potential for benefit sharing MTAs to affect crop genetic resource conservation is unclear. Plant breeders of major crops use germplasm mainly from their own working collections, or acquire it from other breeders, botanists, or geneticists. Typically, this germplasm has already been enhanced and adapted for plant breeding purposes. While exotic germplasm may provide especially useful traits for disease or pest resistance, such germplasm is only one source of the many genes used in an individual variety. Statistics suggest that, for many commercially important crops, only a small percentage of the genes in released varieties are from newly incorporated exotic germplasm (Cox et al., 1988; Goodman and Castillo-Gonzalez, 1991). The expected value of such exotic germplasm is generally small, though on occasion benefits may be larger (Wilkes, 1991). When breeders do require genetic traits unavailable from their conventional sources, gene banks such as the Future Harvest Centers or the NPGS traditionally have had a vast, free supply of germplasm. To date, this germplasm has been provided freely to users, and not subject to MTAs that require benefit sharing. The use of MTAs to market germplasm from some developing countries may also be hindered by a lack of technical expertise. Breeders often require documentation of valuable genetic traits and the ease by which they can be transferred to commercial seed stock. Even if a country has rare and useful germplasm, breeders may remain unaware of its value or existence (Shands, 1994).

To date, the use of MTAs for crop genetic resources has not generated large financial gains for developing countries. In this respect, raw genetic resources, though lacking a well-developed market, are similar to primary export commodities such as timber or coffee. Much of the value added to commercial seed varieties comes from the laborious and time-consuming process of incorporating raw genetic material into elite crop varieties.

Multilateral Agreements Affecting Plant Genetic Resources

Because of the widespread geographic origins and current use of crop genetic resources and the public goods nature of their conservation, the three principal policy tools for conserving genetic resources involve considerable international overlap. A series of multilateral agreements embody the international coordination needed to preserve genetic resources, as well as the lingering debate over property rights for genetic resources.

U.N. Convention on Biological Diversity

The 1993 U.N. Convention on Biological Diversity (CBD) was designed to promote the conservation and sustainable use of biological diversity and to encourage the equitable sharing of resulting benefits. Language in the Convention relating to property rights over genetic materials, biological inventions, technology transfer, and benefit sharing was drafted more with pharmaceutical and industrial development in mind than seed variety development, though subsequent meetings to implement the Convention focused on agricultural biodiversity. On December 29, 1993, the CBD came into force for ratifying and acceding parties (which numbered 188 as of February 15, 2005).[17] Provisions of the Convention have direct implications for the collection, preservation, and exchange of genetic resources. The CBD states that countries have sovereign rights to their indigenous genetic resources, which institutionalizes the change from the practice of freely collecting and sharing of resources. Most countries have interpreted the CBD to allow countries to require payments or transfer of technology in exchange for access to germplasm. The Convention also included a provision for a biosafety protocol to regulate the international movement of the products of biotechnology. Adopted in January 2000, the "Cartagena Protocol" addresses only living modified organisms (LMOs), and makes a distinction between genetically modified organisms as seed and genetically modified organisms intended for food or feed (the assumption being that the latter will not be released into the environment). According to the protocol, LMOs (which include genetically modified seed) are subject to "Advanced Informed Agreement" procedures. Thus, implementation of the protocol has more impact on LMOs that are transferred as seed, or as germplasm for use in genebank system, than on food or feed.

Other agreements play a role. The World Trade Organization (WTO) agreements, which are negotiated, signed, and ratified by the bulk of the world's trading nations, are enforceable through the WTO's ability to levy sanctions. Therefore, countries have strong incentives for the CBD to be consistent with the Trade Related International Property (TRIPS) provisions and the WTO. The International Union for the Protection of New Varieties of Plants (UPOV) is another element affecting the exchange of genetic resources. UPOV-consistent IPR are the leading form of formal varietal protection globally (UPOV protection allows exemptions for breeding and research purposes). After the CBD came into force, the U.S. Department of State (1994) noted that the Convention could not be used to overrule existing intellectual property law, including TRIPS and UPOV. Therefore, both are likely to continue influencing implementation of the CBD.

The International Treaty on Plant Genetic Resources for Food and Agriculture

To address issues left unresolved by the CBD, the International Treaty on Plant Genetic Resources for Food and Agriculture was developed with the intention of (1) mandating conservation of plant genetic resources, (2) ensuring equitable sharing of the benefits created by using these resources, and (3) establishing a multilateral system to facilitate access. The International Treaty entered into force in June 2004 (the U.S. has signed, but not yet ratified the treaty). Sixty-six countries are parties to the treaty. The treaty is to govern international exchange of germplasm and will cover 35 crops, including major cereals like rice, wheat, and maize, but excluding soybean and peanut and other important crops.

IPR have been a major source of debate in interpreting the treaty, particularly the patenting of materials discovered in public gene banks. The treaty states that "Recipients shall not claim any intellectual property or other rights that limit the facilitated access to the plant genetic resources for food and agriculture, [or their genetic parts or components,] [in the form] received from the Multilateral System." Interpretations of this clause abound, particularly with respect to whether the patenting of isolated compounds, such as genes, will be permitted.

The treaty is vague on a number of points. Disagreements remain about the implementation of benefit sharing and the development of a standard Material Transfer Agreement (MTA). The standard MTA is intended to establish the terms of access to plant genetic resources, and all germplasm exchanges under the new multilateral system will be governed by this standard MTA (rather than the bilateral approach suggested by the CBD). The benefits arising from commercial use of germplasm accessed under the multilateral system are to be shared through four mechanisms: (1) exchange of information, (2) access to and transfer of technology, (3) capacity building, and (4) sharing of monetary and other benefits of commercialization. A yet-to-be established portion of monetary benefits from commercial products are to flow, through a trust account managed by the Governing Body of the Treaty, primarily to farmers who conserve genetic resources, especially those in developing and transitional economies.[18] Because benefits will be shared according to conservation practices and income, rather than contributions to the multilateral system, the incentives for conserving genetic resources are likely to be less direct than originally envisioned. More broadly, the means and particulars of financing conservation activities also have not been specified.

Financing International Conservation of Genetic Resources

Given the public good characteristics of crop genetic resources, financing their conservation remains a challenge. Resources available under current and immediately foreseeable policies may be insufficient to conserve the resources agriculture will need. Though MTAs and the expansion of IPR are intended to be self supporting conservation policies, proposals to intensify *in situ* and *ex situ* conservation and to transfer technology and expertise would require additional public funds. Various efforts have been made to estimate actual amounts needed to finance gene banks, *in situ* preservation, and technology transfer. The Keystone International Dialogue (Keystone Center, 1990, 1991) recommended a fund of $300 million annually to support global and national efforts to conserve plant genetic resources. The U.S. National Research Council (1993) recommended that $240 million would be needed annually for maintaining worldwide base collections in addition to evaluation and documentation programs. The FAO (1997) estimated low (A), medium (B), and high (C) funding options ranging from $150 million to $248 million to $455 million annually, averaged over more than ten years. The FAO figures include only costs that would be borne by the international community and do not include domestic program funding. The report considered Option A "basic or rudimentary" while Option B was "consistent with known and documented needs and realistic absorption and implementation capacity of countries" (FAO, 1997).

Grounded in the FAO's Global Plan of Action for genetic resources is a relatively new organization focused more directly on *ex situ* genetic resource preservation. The Global Crop Diversity Trust is an international organization whose establishment has involved a partnership with the FAO and the 16 Future Harvest Centers of the Consultative Group on International Agricultural Research (CGIAR). The Trust aims to match the long-term nature of conservation needs with permanent, sustainable funding by creating an endowment that will perennially fund crop diversity collections around the world. The endowment is intended to facilitate the perpetual conservation of eligible collections that meet agreed standards of management. The Trust will serve as an element of the funding strategy to be implemented under the International Treaty described above.

The Global Crop Diversity Trust hopes to raise a minimum of $260 million from corporations, trusts, foundations, and governments as a permanent endowment for genetic resources. That figure is based on a study carried out by the International Food Policy Research Institute and the University of California, Berkeley, which provided best estimates of the annual funds needed to support the core services provided by the Future Harvest genebanks and the level of endowment needed to provide for the collections in perpetuity (Koo et al., 2002). (The annual costs were estimated to be $5.7 million, and the needed endowment was estimated to be $150 million). The Trust has approximately $45 million in commitments and $70 million under discussion to date (Global Crop Diversity Trust, 2005).

Some researchers have looked at methods beyond multilateral donor systems to fund conservation of genetic resources. Proposals have included a tax on seed sales to provide funds for conservation (Barton and Christensen, 1988). Barton and Christensen suggested either a "straight" sales tax on seed revenues or a system of royalty calculation similar to that used by record companies, with proceeds to be distributed among international, national, and private conservation programs to fund *in situ* and *ex situ* preservation. There are concerns that a royalty based system of direct payments may limit the exchange of genetic resources. Also, if royalty payments in the strict sense are used (i.e., payment upon use in a released variety), returns probably will be limited (Charles, 2001). Proposals to fund germplasm through sales taxes and user fees have been opposed by private seed companies. Even if the proposals were to overcome this opposition, formal seed sales are much less prevalent in self-pollinated crops and in some crops grown in developing countries. Thus, certain crops would not benefit as significantly from this approach.

Another proposal has been to tax agricultural commodities generally (Swaminathan, 1996). This proposal raises questions about the distributional implications (between regions and social classes within regions) of taxing seeds or all agricultural commodities. Because poor families generally spend more on food as a portion of the household budgets, such taxes may be regressive (though to raise equal revenues, the tax rate for a general commodity tax on agricultural, forest, and fish products would need to be only a small fraction of the tax rate for seeds). Another option lies with agricultural producer groups in developed countries (such as Australia, New Zealand, and the United States), many of which fund commodity specific research and market promotion through voluntary checkoff systems that act as a commodity tax. However, while national producer groups may be persuaded to help support domestic gene banks and germplasm characterization, they may be less willing to allocate checkoff funds to an internationally administered fund. As with other aspects of genetic resource use and conservation, private interests do not necessarily coincide with broader public goods.

APPENDIX: MEASURING CROP GENETIC DIVERSITY

Evolutionary or ecological measures of genetic diversity focus particularly on genetic similarity or difference between different species. These kinds of comparisons might also be useful in the study of crop genetic diversity, particularly if a given crop is analyzed in the context of its wild relatives.[19] However, most studies of crop genetic diversity are based on the similarity or difference between different crop populations within the same crop species. Most commonly, named varieties are the crop populations in question, although two distinct varieties may in fact be very similar genetically (Meng *et al.*, 1998). For the rest of this discussion, we will usually assume diversity is being measured within a particular crop species.

Spatial Diversity Measures

Spatial diversity—diversity within a given geographical area—may be "the most commonly recognized concept of diversity" (Meng et al., 1998). Two concepts are often used in spatial measures of genetic diversity. "Richness" refers to a simple count measure, for example of the number of varieties of a particular crop species planted in a given area. "Abundance" is a measure of the evenness of the spatial distribution of elements of the set being considered (Magurran, 1991). For example, suppose the same ten crop varieties are planted in two identical regions. In one region, each variety is planted on one-tenth the area, but in the other region one variety is planted on 91 percent of the area and the other nine varieties occupy one percent each. By a simple count measure (such as richness), the two regions are equally diverse, but introducing abundance would suggest the first region is more diverse than the second. This, along with the fact that named varieties may be very similar genetically, is why simply counting numbers of varieties is likely to be an inadequate measure of crop genetic diversity. Simple diversity indices that reflect varietal distribution (thus partially capturing the concepts of richness and abundance), include the proportion of area planted to the most popular variety or given number of varieties (equivalent to concentration measures used in the industrial organization literature.) A related index is the number of varieties covering a given percentage of total crop area (Widawsky, 1996). Another measure taken from the industrial organization literature is the Herfindahl index, which illustrates the degree of concentration among varieties (Pardey et al., 1996). The Simpson index (one minus the Herfindahl index) and the Shannon-Wiener index, taken from information theory, are often applied in ecological studies of diversity (Magurran, 1991).

Measures of Relationships between Varieties

Other indices of genetic diversity are built up from measures of "genetic distance," i.e., the degree to which varieties or species differ genetically (Nei, 1972; Cavalli-Sforza and Edwards, 1967; Reynolds, Weir, and Cockerham, 1983; Gregorius, 1978). To a certain extent such measures address the problem raised by simply counting named varieties that may be very similar genetically. Genetic distance indices can be calculated based on observations of different crop characteristics, including morphological indicators such as plant height, grain weight, and so on. As indicated, morphological indicators have the advantage that they may be closely linked to the traits on which farmers base their decisions, but the disadvantage that they are often influenced by environment and multiple genes, and therefore not reflective of genetic distance at the chemical (enzyme) or molecular (DNA) level. Genetic distance indices have perhaps most commonly been applied to this biochemical information. The use of biochemical and molecular markers requires systematic physical sampling as well as laboratory time and materials, and as a result can be quite costly (Meng et al., 1998).[20] An alternative approach to measuring genetic distance between varieties, at least for scientifically-bred crops with documented pedigrees, is based on comparison of the heritage of pairs of varieties.[21]

Building Diversity Indices

Genetic distance indices measure differences between different crop varieties or species, but they themselves do not measure overall genetic diversity. Weitzman (1992; 1993) describes a diversity index calculated as the total length of the branches of a taxonomic tree. Such a tree could be calculated using morphological, genealogical (i.e. pedigree), or genetic distance data. Solow, Polasky, and Broadus (1993) also incorporate the size of the set (e.g., number of crop varieties) as well as genetic distance into genetic diversity indices. Both these tree-based measures, and other measures based on matrices of similarity coefficients, permit weighting to reflect the distribution of crop varieties (Souza et al., 1994; Meng et al., 1998).

Measures of Plant Breeding Activity Using Genetic Resources

A number of other measures have been applied to the study of genetic resources, but they usually refer to aspects of a scientific plant breeding program, or the development of such a program from initial crosses involving landraces, rather than to direct measures of genetic diversity. These include numbers and origin of landraces in the ancestry of the varieties being studied, or the number of breeding generations since the initial cross (Gollin and Evenson, 1990); numbers of distinct parental combinations and numbers of unique landrace ancestors per pedigree (Smale and McBride, 1996; Hartell, 1996; Smale et al., 1998); or coefficient of parentage (COP) based measures (Pardey et al., 1996). Note that all of these pedigree-based measures are less useful in a crop, such as corn, that may not always follow a strict pedigree breeding system, or in crops for which pedigrees are partially or completely private for proprietary reasons.

Temporal Diversity

Duvick (1984) observed that in a number of scientifically-bred crops, *temporal* diversity (or diversity through time) has replaced spatial diversity as one means of maintaining or even raising resistance or tolerance to pests and diseases. Temporal diversity depends on maintaining breeding effort by humans. Meng et al. (1998) closely identify temporal diversity with "the rate of change or turnover of [planted] varieties" as defined, for example, by Brennan (1984) and Brennan and Byerlee (1991). Other things being equal, faster varietal turnover might be expected to be associated with increased temporal genetic diversity, but like pedigree-based measures, varietal turnover is more a measure of the output of a plant breeding program than of genetic diversity *per se*. Newly released varieties might be genetically somewhat dissimilar to older varieties, or they might be very closely related genetically. Time-series of *spatial* diversity measures could provide useful information about *temporal* change in diversity, but such a series would not strictly measure "temporal diversity." More formal assessment of temporal genetic diversity could be made by statistically testing differences between genetic distance measures over temporal samples (See Souza et al., 1994 and Tessier and Bernatchez, 1999).

REFERENCES

Acharya, R. (1991). "Patenting of Biotechnology, GATT and the Erosion of the World's Biodiversity," *Journal of World Trade,* Vol. 25,71-87.

Ahearn, M., Yee, J., Ball, E. & Nehring, R. (1998). *Agricultural Productivity in the United States.* AIB-740. Economic Research Service, United States Department of Agriculture.

Alston, J. M., Norton, G. W. & Pardey, P. G. (1995). *Science Under Scarcity: Principles and Practice for Agricultural Research Evaluation and Priority Setting*. Ithaca, New York: Cornell University Press.

Attwood, T. K. (2000). "The Babel of Bioinformatics." *Science, 290*, 471-473.

Barbier, E. B., Brown, G., Dalmozzone, S., Folke, C., Gadgil, M., Hanley, N., Holling, C. S., Lesser, W. H., Maler, K. G., Mason, P., Panayotou, T., Perrings, C., Turner, R. K. & Wells, M. (1995). "The Economic Value of Biodiversity," *Global Biodiversity Assessment,* V. H., Heywood (ed.). Cambridge and New York: Cambridge University Press for the United Nations Environment Program.

Barton, J. H. & Christensen, E. (1988). "Diversity Compensation Systems: Ways to Compensate Developing Nations for Providing Genetic Material," *Seeds and Sovereignty: The Use and Control of Plant Genetic Resources*, J. R. Kloppenburg (ed.). Durham, NC: Duke University Press.

Barton, J. & Siebeck, W. (1991). *Intellectual Property Issues for the International Agricultural Research Centers: What are the Options?* Report prepared for the CGIAR Secretariat, Washington, DC.

Blum, E. (1993). "Making Biodiversity Conservation Profitable: A Case Study of the Merck/INbio Agreement," *Environment*, Vol. *35*, No. 4.

Brennan, J., Singh, I. P. & Lewin, L. G. (1997). "Identifying International Rice Research Spillovers in New South Wales," *Agricultural Economics*, Vol. *17*.

Brennan, J. P. (1984). "Measuring the Contribution of New Varieties to Increasing Wheat Yields," *Review of Marketing and Economics*, Vol. 52.

Brennan, J. P. & Byerlee, D. (1991). "The Rate of Crop Varietal Replacement on Farms: Measures and Empirical Results for Wheat," *Plant Varieties and Seeds*, Vol. *4*.

Brown, G. (1987). "Valuation of Genetic Resources" *The Preservation and Valuation of Biological Resources*, G. Orians, G. Brown, W. Kunin, J. Swierzbinski (eds.). Seattle. WA: University of Washington Press.

Brown, G. & Swierzbinski, J. (1985). "Endangered Species, Genetic Capital, and Cost-Reducing R&D," *Economics of Ecosystem Management,* D. Hall, N. Myers, and N. Margaris (eds.).

Bruinsma, J. (ed.) 2003. *World Agriculture: Towards 2015/2030. An FAO Perspective.* London: Earthscan Publications, Ltd.

Brush, S. B. (1992). "Farmers' Rights and Genetic Conservation in Traditional Farming Systems," *World Development*, Vol. *20*, No. 11.

Brush, S. B. (1994). *Providing Farmers'Rights through In Situ Conservation of Crop Genetic Resources.* Report to the FAO Commission on Plant Genetic Resources. Rome: U.N. Food and Agriculture Organization.

Byerlee, D. & Traxler, G. (1995). "National and International Wheat Improvement Research in the Post Green Revolution Period: Evolution and Impacts." *American Journal of Agricultural Economics*, Vol. *77*.

Cabanilla, V. L., Hossain, M. & Khush, G. S. (1999). "*Diffusion of Breeding Materials and Genetic Composition of Improved Rices in South and Southeast Asia 1965-98.*" Paper presented at the Impacts Assessment and Evaluation Group (IAEG) Germplasm Research Impact Study Workshop, Nashville, TN, Aug. 7-8.

Cavalli-Sforza, L. L. & Edwards, A. W. F. (1967). "Phylogenetic Analysis: Models and Estimation Procedures.." *American Journal of Human Genetics*, Vol. *19*.

Chang, T. T. (1994). "The biodiversity crisis in Asia crop production and remedial measures," *Biodiversity and Terrestrial Ecosystems*.C.I. Peng and C.H. Chou (eds.), Institute of Botany, Academia Sinica Monograph Series No. 14. Taipei: Academia Sinica.

Charles, D. (2001). "Seeds of Discontent," *Science*, Vol. *294* (October).

Christensen, E. (1987). "Genetic Ark: A Proposal to Preserve Genetic Diversity for Future Generations," *Stanford Law Review,* Vol. *40*.

Cooper, J. C. (1998). "The Economics of Public Investment in Agro-biodiversity Conservation," *gricultural Values of Plant Genetic Resources*. In R.E. Evenson, D. Gollin, and V. Santaniello (eds.), Wallingford, UK:

CABI Publishing.

Cox, T., Murphy, J. & Goodman, M. (1988). "The Contribution of Exotic Germplasm to American Agriculture," *Seeds and Sovereignty: The Use and Control of Plant Genetic Resources*. J.R. Kloppenburg (ed.). Durham, NC: Duke University Press.

Cox, T. S., Kiang, Y. T., Gorman, M. B. & Rodgers, D. M. (1985). "Relationship Between Coefficient of Parentage and Genetic Similarity Indices in the Soybean," *Crop Science*, Vol. *25*.

Day, P. R. (1997). "Biodiversity and the Equitable Use of the World's Genetic Resources," *Global Genetic Resources: Access, Ownership and Intellectual Property Rights,* K. E. Hoagland and A. Y. Rossman (eds.) Beltsville symposia in Agricultural Research. Washington, DC: The Association of Systematics Collections.

Day-Rubenstein, K., Stuart, M. & Frisvold, G. (2000). "Agricultural land use in tropical forested countries: patterns, determinants, and implications for biodiversity loss," *World Resource Review*, Vol. *12 (4)*.

Duvick, D. N. (1977). "Genetic Rates of Gain in Hybrid Maize Yields During the Past 40 Years," *Maydica*, Vol. *22 (4)*.

Duvick, D. N. (1984). "Genetic Diversity in Major Farm Crops on the Farm and in Reserve," *Economic Botany*, Vol. *38*.

Duvick, D. N. (1986). "Plant Breeding: Past Achievements and Expectations for the Future," *Economic Botany*, Vol. *40*, No. 3.

Duvick, D. N. (1991). "Industry and Its Role in Plant Diversity," *Forum for Applied Research and Public Policy*, Vol. *6*, No. 3.

Duvick, D. N. (1992). "Plant Breeding: Past Achievements and Expectations for the Future," *Economic Botany*. Vol. *40*.

Duvick, D. N. (1998). "The United States," *Maize Seed Industries in Developing Countries*. M.L. Morris (ed.), Boulder, CO: Lynne Rienner Publishers.

Evenson, R. & D. Gollin (1997). "Genetic Resources, International Organizations and Improvement in Rice Varieties," *Economic Development and Cultural Change,* Vol. *45*.

Evenson, R. E. & Lemarié, S. (1998). "Optimal collection and search for crop genetic resources," *Farmers, Gene Banks and Crop Breeding: Economic Analyses of Diversity in Wheat, Maize and Rice*, M. Smale (ed.), London: Kluwer.

Falck Zepeda, J. & Traxler, G. (1997). "*The Role of Federal, State and Private Institutions in Seed Technology Generation in the United States: The Case of Cotton.*" Paper presented at the Economic Research Service/Farm

Foundation Workshop on Agricultural R&D, Rosslyn, VA, Aug. 5.

Feyerherm, A. M. & Paulsen, G. M. (1981). "An Analysis of Temporal and Regional Variation in Wheat Yields," *Agronomy Journal*, Vol. *73*.

Feyerherm, A. M., Paulsen, G. M. & Sebaugh, J. L. (1984). "Contribution of Genetic Improvement to Recent Wheat Yield Increases in the USA," *Agronomy Journal*, Vol. *76*.

Fowler, C. (1991). "Stakes High in Battle for Genetic Diversity," *Forum for Applied Research and Public Policy,* Vol. *6*, No. 3.

Frankel, O. H. (1970). "The genetic dangers of the Green Revolution," *World Agriculture*, Vol. *19*, No. 3.

Frisvold, G. & Condon, P. (1994). "Biodiversity Conservation and Biodiversity Development Agreements," *Contemporary Economic Policy*, Vol. *12*.

Frisvold, G., Sullivan, J. & Raneses, A. (2003). "Genetic improvements in major U.S. crops: the size and distribution of benefits," *Agricultural Economics*, Vol. *28*, No. 2 (March).

Fuglie, K., Ballenger, N., Day, K., Klotz, C., Ollinger, M., Reilly, J., Vasavada, U. & Yee, J. (1996). *Agricultural Research and Development: Public and Private Investments Under Alternative Markets and Institutions,* AER-735, U.S. Department of Agriculture, Economic Research Service.

Global Crop Diversity Trust (2005) www.startwithaseed.org/items/ trust.php Accessed March 11, 2005.

Gollin, D. & Evenson, R. (1990). *Genetic Resources and Rice Varietal Improvement in India*, New Haven, CT: Economic Growth Center, Yale University, unpublished manuscript.

Gollin, D., Smale, M. & Skovmand, B. (2000). "Searching an *Ex Situ* Collection of Genetic Resources," *American Journal of Agricultural Economics*, Vol. *82*.

Goodman, M. & Castillo-Gonzalez, F. (1991). "Plant Genetics: Politics and Realities," *Forum for Applied Research and Public Policy*, Vol. *6*.

Gregorius, H. R. (1978). "The Concept of Genetic Diversity and its Formal Relationship to Heterozygosity and Genetic Distance," *Mathematical Biosciences*, Vol. *41*.

Hanemann, W. M. (1988). "Economics and the Preservation of Biodiversity," *Biodiversity*, E. O. Wilson (ed.), Washington, DC: National Academy Press.

Harlan, J. R. (1971). "Agricultural Origins: Centers and Noncenters," *Science*, Vol. *174*.

Harlan, J. R. (1992). *Crops and Man.* Madison, WI: American Society of Agronomy, Inc., and Crop Science Society of America, Inc.

Hartell, J. G. (1996). *The Contribution of Genetic Resource Diversity: The Case of Wheat Productivity in the Punjab of Pakistan*, M.Sc. Thesis, University of Minnesota, St. Paul, MN.

Heisey, P. W., Srinivasan, C. S. & Thirtle, C. (2001). *Public Sector Plant Breeding in a Privatizing World.* AIB-772. U.S. Department of Agriculture, Economic Research Service.

Heisey, P. W., Lantican, M. A. & Dubin, H. J. (2002). *Impacts of International Wheat Breeding Research in Developing Countries, 1966-97*. Mexico, D.F.: CIMMYT.

Houghton, R. A. (1994). "The Worldwide Extent of Land-use Change," *BioScience*, Vol. *44*, No. 5.

Kaplan, J. K. (1998). "Conserving the World's Plants," *Agricultural Research*, Vol. *46*, no. 9.

Kempthorne, O. (1969). *An Introduction to Genetic Statistics.* Ames, IA: Iowa State University.

Keystone Center (1990). *Final Consensus Report of the Keystone International Dialogue Series on Plant Genetic Resources, Madras Plenary Session.* Washington, DC: Genetic Resources Communications Systems.

Keystone Center (1991). *Final Consensus Report: Global Initiative for the Security and Sustainable Use of Plant Genetic Resources: Keystone International Dialogue Series on Plant Genetic Resources, Oslo Plenary Session.* Washington, DC: Genetic Resources Communications Systems.

Kloppenburg, J. R. (1988). *First the Seed: The Political Economy of Plant Biotechnology*, Cambridge: Cambridge University Press.

Knudson, M. (1998). "Agricultural Diversity: Do We Have the Resources to Meet Future Needs?" *Global Environmental Change: Assessing the Impacts*. G. Frisvold and B. Kuhn (eds.). Aldershot: Edward Elgar Publishing.

Koo, B., Pardey, P. G. & Wright, B. D. with Bramel, P., Debouck, D., van Dusen, M. E., Jackson, M. T., Rao, N. K., Skovmand, B., Taba, S. & Valkoun, J. (2004). *Saving Seeds: The Economics of Conserving Genetic Resources Ex Situ in the Future Harvest Centres of the CGIAR*. Wallingford, UK: CABI Publishing.

Koo, B., Pardey, P. G. & Wright, B. D. (2002). *Endowing Future Harvests: The Long-Term Costs of Conserving Genetic Resources at the CGIAR* Centres, International Plant Genetic Resources Institute, Rome.

Kronstad, W. E. (1996). "Genetic Diversity and the Free Exchange of Germplasm in Breaking Yield Barriers," *Increasing Yield Potential in Wheat: Breaking the Barriers*. M.P. Reynolds, S. Rajaram, and A. McNab (eds.), Mexico, D.F.: CIMMYT.

Magurran, A. (1991). *Ecological Diversity and Its Measurement*. Princeton, NJ: Princeton University Press.

Malecot, G. (1948). *Les Mathematiques de l'Hérédité*. Paris: Masson.

Meng, E.C.H., M. Smale, M. Bellon and D. Grimanelli (1998). "Definition and Measurement of Crop Diversity for Economic Analysis," *Farmers, Gene Banks and Crop Breeding: Economic Analyses of Diversity in Wheat, Maize, and Rice*. M. Smale (ed.) Boston, MA: Kluwer Academic Publishers.

Mooney, P. (1979). *Seeds of the Earth: a Private or Public Resource?* Ottawa: Canadian Council for International Cooperation.

Mooney, P. (1983). "The Law of the Seed: Another Development and Plant Genetic Resources," *Development Dialogue*, Vol. *12*.

Morris, M. L. (2002). *Impacts of International Maize Breeding Research in Developing Countries*, 1966-98. Mexico, D.F.: CIMMYT.

Morris, M. L. & Heisey, P. W. (2003). "Estimating the benefits of plant breeding research: methodological issues and practical challenges," *Agricultural Economics, 29*.

Morris, M. L., Risopoulos, J. & Beck, D. (1999). *Genetic Change in Farmer-Recycled Maize Seed: A Review of the Evidence*. CIMMYT Economics Program Working Paper 99-07. Mexico, D.F.: CIMMYT.

National Research Council (NRC) (1972). *Genetic Vulnerability of Major Crops*. Washington, DC: National Academy of Sciences.

National Research Council (NRC) (1991). *Managing Global Genetic Resources: The U.S. National Plant Germplasm System*, Committee on Managing Global Genetic Resources: Agricultural Imperatives, Board on Agriculture, National Research Council. Washington, DC: National Academy Press.

National Research Council (NRC) (1993). *Managing Global Genetic Resources: Agricultural Crop Issues and Policies*. Committee on Managing Global Genetic Resources: Agricultural Imperatives, Board on Agriculture, National Research Council. Washington, DC: National Academy Press.

Nei, M. (1972). "Genetic Distance Between Populations," *American Naturalist*, Vol. *106*, pp. 283-292.

Nightingale, K. J. (1996). *Trends in Genetic Diversity of Spring Wheat in the*

Developing World. East Lansing, MI: M.Sc. Thesis, Michigan State University.

Office of Technology Assessment, United States Congress (OTA) (1987). *Technologies to Maintain Biological Diversity*, OTA-F-330.

Pardey, P. G., Alston, J. M., Christian, J. E. & Fan, S. (1996). *Hidden Harvest: U.S. Benefits from International Research Aid.* Food Policy Report. Washington, DC: International Food Policy Research Institute.

Pardey, P. G., Koo, B., Wright, B. D., van Dusen, M. E., Skovmand, B. & Taba, S. (2001). "Costing the Conservation of Genetic Resources: CIMMYT's *Ex Situ* Maize and Wheat Collection," *Crop Science* Vol. *41*, No. 4.

Pardey, P. G., Koo, B., van Dusen, M. E., Skovmand, B., Taba, S. & Wright, B. D. (2004). "CIMMYT Genebank," *Saving Seeds: The Economics of Conserving Genetic Resources Ex Situ in the Future Harvest Centres of the CGIAR.* Wallingford, UK: CABI Publishing.

Plucknett, D., Smith, N. J. H., Williams, J. & Anishetty, N. (1987). *Genebanks and the World's Food.* Princeton, NJ: Princeton University Press.

Porceddu, E., Ceoloni, C., Lafiandra, D., Tanzarella, O. A. & Scarascia Mugnozza, G. T. (1988). "Genetic resources and plant breeding: problems and prospects," *The Plant Breeding International, Cambridge Special Lecture*. Cambridge, UK: Institute of Plant Science Research.

Pray, C. E. (1990). "Intellectual Property Rights on Plants: Enforcement, R&D and Potential Importance for LDCs." Mimeo. Rutgers University, New Brunswick, NJ.

Pray, C. E. & Knudson, M. (1994). "Impact of Intellectual Property Rights on Genetic Diversity: A Case Study of U.S. Wheat," *Contemporary Policy Issues* (Jan.).

Pray, C. E., Knudson, M. & Masse, L. (1993). "Impact of Changing Intellectual Property Rights on U.S. Plant Breeding R&D." Paper presented at NC204 Conference on U.S. Agricultural Research. Orlando, FL. March.

Rausser, G. & Small, A. (2000). "Valuing research leads: bioprospecting and the conservation of genetic resources," *Journal of Political Economy,* Vol. *108*, No. 1.

Reid, W. R. (1992). *Genetic Resources and Sustainable Agriculture: Creating Incentives for Local Innovation and Adaptation.* Biopolicy International Series No. 2. Nairobi, Kenya: ACTS Press.

Rejesus, R., van Ginkel, M. & Smale, M. (1996). "Wheat Breeders' Perspectives on Genetic Diversity and Germplasm Use: Findings from an

International Survey," *Plant Varieties and Seeds,* Vol. *9*, No. 3.

Reynolds, J., Weir, B. S. & Cockerham, C. C. (1983). "Estimation of the Coancestry Coefficient: Basis for a Short-Term Genetic Distance," *Genetics* Vol. *105*.

Schmidt, J. W. (1984). "Genetic Contributions to Yield Gains in Wheat," *Proceedings of a Symposium on Genetic Contributions to Yield Gains of Five Major Crop Plants*, W.R. Fehr (ed.). Sponsored by C-1, Crop Science Society of America, Atlanta, GA, Dec. 2, 1981. Crop Science Society of America and American Society of Agronomy, Madison, WI.

Sedjo, R. (1992). "Property Rights, Genetic Resources, and Biotechnological Change," *Journal of Law and Economics*, Vol. *35*.

Shands, H. L. (1994). "Some Potential Impacts of the Convention on Biological Diversity on the International System of Exchange of Food Crop Germplasm," *Conservation of Plant Genetic Resources and the U.N. Convention on Biological Diversity*, D. Witmeyer and M. Strauss (eds.). Washington, DC: American Association for the Advancement of Science.

Shands, H. L. & Stoner, A. K. (1997). "Agricultural Germplasm and Global Contributions," in *Global Genetic Resources: Access, Ownership and Intellectual Property Rights,* K. E. Hoagland and A. Y. Rossman (eds.). Beltsville Symposia on Agricultural Research, Washington, DC: The Association of Systematics Collections.

Siebeck, W. E. (1994). *Personal communication*. Special Advisor to the Vice President for Development Economics, The World Bank, Washington, DC.

Simpson, D. R. & Sedjo, R. A. (1998). "The value of genetic resources for use in agricultural improvement," *Agricultural Values of Plant Genetic Resources,* R.E. Evenson, D. Gollin and V. Santaniello (eds.). Wallingford, UK, CABI Publishing.

Simpson, D. R., Sedjo, R. A. & Reid, J. W. (1996). "Valuing Biodiversity for Use in Pharmaceutical Research," *Journal of Political Economy*, Vol. *104*, No. 1.

Simpson, D. R. & Sedjo, R. A. (1992). "Contracts for Transferring Rights to Indigenous Genetic Resources," *Resources*, Fall.

Smale, M. (1997). "The Green Revolution and Wheat Genetic Diversity: Some Unfounded Assumptions," *World Development*, Vol. *25*, No. 8.

Smale, M. (ed.) (1998). *Farmers, Gene Banks and Crop Breeding: Economic Analyses of Diversity in Wheat, Maize, and Rice.* Boston, MA: Kluwer Academic Publishers.

Smale, M., Reynolds, M. P., Warburton, M., Skovmand, B., Trethowan, R.,

Singh, R. P., Ortiz-Monasterio, I., Crossa, J., Khairallah, M. & Almanza-Pinzon, M. I. (2001). *Dimensions of Diversity in CIMMYT Bread Wheat from 1965 to 2000*. Mexico, D.F.: CIMMYT.

Smale, M. & McBride, T. (1996). "Understanding Global Trends in the Use of Wheat Diversity and International Flows of Wheat Genetic Diversity," Part 1 of *CIMMYT 1995/6 World Wheat Facts and Trends 1995/6: Understanding Global Trends in the Use of Wheat Diversity and International Flows of Wheat Genetic Diversity*. Mexico, D.F.: CIMMYT.

Smale, M., Hartell, J., Heisey, P. & Senauer, B. (1998). "The Contributions of Genetic Resources and Diversity to Wheat Production in the Punjab of Pakistan," *American Journal of Agricultural Economics*, Vol. *80*.

Smale, M. & Day-Rubenstein, K. (2002). "The Demand for Crop Genetic Resources: International Use of the U. S. National Plant Germplasm System," *World Development*, Vol. *30*, No. 9.

Solow, A., Polasky, S. & Broadus, J. (1993). "On the Measurement of Biological Diversity," *Journal of Environmental Economics and Management*, Vol. *24*.

Souza, E., Fox, P., Byerlee, D. & Skovmand, B. (1994). "Spring Wheat Diversity in Irrigated Areas of Two Developing Countries," *Crop Science*, Vol. *34*.

Sperling, C. (1994). Personal communication. United States Department of Agriculture, National Germplasm Resources Laboratory, Beltsville, MD.

Swaminathan, M. (1996). "Compensating Farmers and Communities through a Global Fund for Biodiversity Conservation for Sustainable Food Security," *Diversity*, Vol. *12*.

Swanson, T. (1996). "Global Values of Biological Diversity: The Public Interest in the Conservation of Plant Genetic Resources of Agriculture," *Plant Genetic Resources Newsletter No. 105*. Rome: International Plant Genetic Resources Institute.

Tessier, N. & Bernatchez, L. (1999). "Stability of Population Structure and Genetic Diversity Across Generations Assessed by Microsatellites Among Sympatric Populations of Landlocked Atlantic Salmon (*Salmo salar* L.)," *Molecular Ecology*, Vol. *8*.

Thirtle, C. G. (1985). "Technological Change and the Productivity Slowdown in Field Crops: United States, 1939-78," *Southern Journal of Agricultural Economics*, Vol. *17* (Dec.).

Tilman, D. (1998). "The greening of the Green Revolution," *Nature*, Vol. *396*.

U.N. FAO (1996a). *Report of the International Technical Conference on Plant Genetic Resources*. Rome.

U.N. FAO (1996b). *Report on the State of the World's Plant Genetic Resources*. Rome.

U.N. FAO (1996c). *Global Plan of Action for the Conservation and Sustainable Utilization of Plant Genetic Resources for Food and Agriculture, and the Leipzig Declaration*. Adopted by the International Technical Conference on Plant Genetic Resources, Leipzig, Germany, June 17-23, Rome.

U.N. FAO (1997). *Revision of the Cost Estimates of the Global Plan of Action*. Rome.

U.N. FAO (1998). *The State of the World's Plant Genetic Resources for Food and Agriculture*. Rome.

U.S. Department of Agriculture (1990). "New Germplasm Helps Assure Food for the Future," USDA News Feature, News Division, Office of Public Affairs, Dec. 6.

U.S. Department of Agriculture, Economic Research Service (2002). ERS-USDA Land Use Database. www.ers.usda.gov/Briefing/LandUse/agchangechapter.htm#regional

United States Department of State (1994). Correspondence between Warren Christopher, Secretary of State, and Senator Robert Dole, Aug. 16.

United States General Accounting Office (GAO) (1997). *U. S. Department of Agriculture: Information on the Condition of the National Plant Germplasm System*. Report to Congressional Committees.

Webster, P. (2003). "Prestigious Plant Institute in Jeopardy," *Science,* Vol. *229* (Jan. 31).

Weitzman, M. L. (1992). "On Diversity," *Quarterly Journal of Economics* Vol. *107*.

Weitzman, M. L. (1993). "What to Preserve? An Application of Diversity Theory to Crane Conservation," *Quarterly Journal of Economics*, Vol. *108*.

Widawsky, D. A. (1996). *Rice Yields, Production Variability, and the War Against Pests: An Empirical Investigation of Pesticides, Host-Plant Resistance, and Varietal Diversity in Eastern China*. Palo Alto, CA: Ph.D. dissertation, Stanford University.

Wilkes, G. (1991). "*In Situ* Conservation of Agricultural Systems," *Biodiversity: Culture, Conservation, and Ecodevelopment,* J. B. Alcorn and M. Oldfield, (eds.). Boulder, CO: Westview Press.

World Resources Institute (WRI), United Nations Environment Programme, United Nations Development Programme, and The World Bank (1991). *World Resources, 1990-1991*. New York: Oxford University Press.

Wright, S. (1922). "Coefficients of Inbreeding and Relationship," *American Naturalist,* Vol. *56*.

Zohrabian, A. (1995). *Genetic Resources Conservation and Maintenance in the Former Soviet Union.* Department of Agricultural and Applied Economics Staff Paper P951. Minneapolis: University of Minnesota.

End Notes

[1] Genetic resources may also have economic value even if they are not currently being used. By preserving resources, we retain the option to use them in the future, when they may become important for agricultural, pharmaceutical, ecological, or industrial applications—even if we do not currently know precisely what those resources or applications are (Kaplan, 1998). Even if they are never used, diverse genetic resources may be valued by some people simply for their existence, or as a bequest left intact to future generations (Barbier et al., 1995).

[2] Growth accounting is often indicative rather than exact. Various factors (such as improved germplasm and improved crop management practices) frequently interact with one another, making it difficult to isolate the contributions of a single source. Interaction also means that the productivity gain from simultaneous adoption often exceeds the sum of the productivity gains when new varieties or crop management practices are adopted separately (Morris and Heisey, 2003).

[3] See Duvick (1977, 1984, 1992) on maize (corn) in the U.S., and Feyerherm and Paulsen (1981); Feyerherm, Paulsen, and Sebaugh (1984); Schmidt (1984); Cox et al. (1988); and others listed by Heisey, Lantican, and Dubin (2002) on wheat in both industrialized and developing countries.

[4] Most studies have focused only on genetic improvements for a single crop.

[5] The average annual growth in U.S. crop yields during 1975–92 was 1.33 percent for corn, 1.54 percent for sorghum, 1.13 percent for wheat, 1.23 percent for soybeans, and 2.23 percent for cotton. The half of the yield growth not attributed to improved seed varieties came from other inputs and management factors, including more fertilizers and pesticides, better agronomic practices, and investments in irrigation and drainage. These other sources of productivity growth also may have been affected by agricultural research.

[6] In other words, annual yield gains attained in a given year are maintained in the years following.

[7] Of course in the long term, research gains may be counteracted by losses of resistance to pests and diseases, but in a successful research program the net gains are positive. The point here is that it is more realistic to look at research gains as a permanent stream over time rather than as an economic benefit occurring only once. If investments are ongoing, new additions to the permanent stream are received every year. Furthermore, avoidance of losses is in fact an economic benefit as well.

[8] Koo et al. present additional cost figures for CGIAR gene banks. For many cost compoinents of gene bank operation, cost estimates fall between the estimates for wheat and maize.

[9] For example, certain isolated areas in Mediterranean Europe grow wheat landraces. Faro, or Triticum dicoccum, is grown in Italy.

[10] This refers to countries for which a large proportion of the corn produced enters the formal market.

[11] In another example, modern corn hybrids adapted to the Midwestern United States were derived from dent varieties from the Southeastern United States and flint varieties from the Northeast, which were themselves adapted by settler farmers from many locally distinct

varieties selected and reselected by Native American farmers over many previous generations (Duvick, 1998)

[12] After the breakup of the Soviet Union, the Vavilov Institute, one of the largest collections in the world, has faced critical financial and structural problems. Funding for gene banks in Russia and in these republics has been greatly reduced and many accessions are at risk (Zohrabian, 1995; Webster, 2003).

[13] A GAO study of the U.S. Plant Germplasm System echoed the concerns of the FAO report (GAO, 1997).

[14] Goodman and Castillo-Gonzalez (1991) also note that "improved breeding lines have been less freely exchanged, even among public agencies."

[15] Unshielded genetic materials also contain improved varieties; in fact, improved breeding materials are the type of germplasm most frequently distributed by the U.S. National Plant Germplasm System. While public research institutions are the primary source of germplasm placed in the NPGS, private breeding concerns donate materials as well, particularly obsolete breeding materials.

[16] Utility patents are the broadest class of patents and, unlike plant patents, they can be used for sexually reproducing plants.

[17] The United States signed the Convention in June 1993, but the U.S. Senate has not yet ratified it

[18] Other aspects of the MTAs, such as recordkeeping and means of assigning parentage of a variety, have yet to be worked out in detail.

[19] See Smale (1998), and particularly the chapter by Meng et al. (1998), for one of the first attempts to summarize the application of various diversity- related measures to crops and to give these measures an economic interpretation.

[20] Another characteristic, infrequently noted, of both morphological and genetic measures is that they obviously require informed choice of the characteristics or genes that will be analyzed. No index will be constructed, for example, based on all genes in a crop that are polymorphic, i.e., genes that have more than one variant. In the first place, such a list is unknown, and in the second, costs would become completely prohibitive.

[21] This approach uses the coefficient of diversity (COD), which equals 1 - the coefficient of parentage (COP). The COP is a pairwise comparison based on pedigree analysis (Wright, 1922; Malecot, 1948; Kempthorne, 1969; Cox et al., 1985). COD/COP analysis is less costly than analysis of proteins or molecular methods, but it also has some disadvantages: 1) it ignores the possibility that alleles could be identical even without common heritage; 2) it relies on the assumption that the ultimate ancestors that are recorded in a pedigree are unrelated, which may not be true; and 3) it assumes that "each parent contributes equally to offspring, despite the effects of recurrent selection and random genetic drift" (See Nightingale, 1996; Cox et al., 1985; Meng et al., 1998)

In: Crop Genetic Resources
Editor: Sean W. Price
ISBN: 978-1-63483-135-2

Chapter 3

THE SEED INDUSTRY IN U.S. AGRICULTURE: AN EXPLORATION OF DATA AND INFORMATION ON CROP SEED MARKETS, REGULATION, INDUSTRY STRUCTURE, AND RESEARCH AND DEVELOPMENT*

Jorge Fernandez-Cornejo

ABSTRACT

The unprecedented growth in crop yields and agricultural total factor productivity over the past 70 years owes much to a series of biological innovations embodied in seeds, beginning with the development of hybrid crops in the United States in the early part of the 20th century, continuing with the adoption of high-yielding varieties during the Green Revolution of the 1960s and 1970s, and more recently, modern biotechnology. Throughout this period, the seed industry evolved, as small businesses gave way to larger enterprises that integrated plant breeding, production, conditioning, and marketing functions. The industry was further shaped by widespread mergers and acquisitions in the latter part of the century, rapid growth in private research and

* This is an edited, reformatted and augmented version of Agriculture Information Bulletin Number 786, issued by the U.S. Department of Agriculture, Economic Research Service, January 2004.

development (R&D), shifting roles of public and private R&D, and a "coming of age" of agricultural biotechnology.

Keywords: Seed markets, seed costs and prices, regulation, plant breeding, field crops, research and development, industry concentration, biotechnology.

SUMMARY

The U.S. seed industry changed dramatically over the past century, as more farmers purchased their seed (instead of using seed saved from the previous harvest) and small seed businesses gave way to larger enterprises that integrated plant breeding, production, conditioning, and marketing functions. The industry was further shaped by widespread mergers and acquisitions in the latter part of the 20th century, rapid growth in private research and development (R&D), shifting roles of public and private R&D, and a "coming of age" of agricultural biotechnology.

To assess these developments, this report examines the composition of U.S. and international seed markets, regulations affecting agricultural seeds, the structure and evolution of the seed industry, and trends in private and public R&D in plant breeding. Particular emphasis is placed on seeds for the major field crops: corn, cotton, soybeans, and wheat.

Improved Seed Is a Major Contributor to Crop Yield Gains

Over the past 70 years, yields of all major field crops in the United States registered a remarkable increase. For example, average corn yields rose from 20 bushels per acre in 1930 to 140 bushels per acre by the mid-1990s. Over the same period, cotton yields rose nearly fourfold, soybean yields increased more than threefold, and wheat yields climbed more than 2.5-fold. More than half of the yield gains are attributed to genetic improvements achieved by plant breeders.

Purchased Seed Use, Purchase Value, and Trade Have Grown in Recent Decades

The United States is the largest seed market worldwide, followed by China and Japan. Seed expenditures by U.S. farmers rose from about $500 million per year in 1960 to nearly $7 billion in 1997. In real terms, seed expenditures climbed about 2.5-fold in the same period, despite minimal real increases in the index of seed prices paid by farmers. A large portion of the increase in real seed expenditures may be explained by increases in the share of seed purchased by U.S. farmers from commercial sources, which, in turn, can be explained by increases in seed productivity attributable to scientific improvements in plant breeding. The United States is a net exporter of seed. In 1996, the U.S. seed trade surplus was $384 million: $698 million in seed exports, mainly to Mexico, Canada, Italy, Japan, and Argentina; and $314 million in seed imports, mainly from Canada, Chile, the Netherlands, and China.

Intellectual Property Rights Effect Significant Changes

Hybrid corn varieties developed in the first half of the 20th century and, widely accepted by farmers, provided the private sector a natural method of protecting plant breeding investments—saved hybrid corn seed produces substantially lower yields, encouraging farmers to repurchase seed every year. This development, combined with a strengthening of legal protection of intellectual property rights in the second half of the 20th century, brought large-scale change to the seed industry, particularly increases in R&D and industry concentration.

Seed Industry Structure Is Characterized by Growth and Consolidation

From the commercial production of hybrid corn seed in the 1930s to the recent mergers and acquisitions, the history of the U.S. seed industry is marked by extensive structural change and transition. Until the 1930s, most commercial seed suppliers were small, family-owned businesses lacking the financial resources necessary to pursue their own research and development. These small businesses depended almost exclusively on plant breeding research in the public sector. Seed businesses served primarily to multiply and sell seeds of varieties developed in the public domain.

Market concentration in the U.S. seed industry increased in the latter part of the 20th century, with the four largest corn seed firms accounting for nearly 70 percent of U.S. corn seed sales in 1997 and the four largest cotton seed firms providing more than 90 percent of the cotton seed varieties planted. In contrast, the public sector still accounts for a large share of the wheat seed varieties used by U.S. farmers. Although the increase in seed industry concentration has raised concerns about its potential impact on market power, preliminary empirical results for U.S. cotton and corn seed industries over the past 30 years suggest that increased concentration resulted in a cost-reducing effect that prevailed over the effect of enhanced market power.

Private and Public Sector Roles in Crop Seed R&D Have Shifted

Private R&D expenditures on plant breeding increased 1,300 percent between 1960 and 1996 (adjusted for inflation), while real public R&D expenditures changed little. With the development of commercially viable corn hybrids in the 1930s, R&D expenditures on corn varieties were the first to shift from public to private. Private seed companies' share of total expenditures on plant breeding R&D on corn increased from close to half in 1970 to more than 70 percent in 1989. The shift from public to private R&D expenditures on soybean plant breeding is more recent, as the share of private sector R&D for soybeans rose from 6 percent to almost 25 percent between 1970 and 1984. Private sector R&D for improved wheat varieties has been limited. As a result, farmers have relied on public sector wheat varieties for new sources of seed. Public sector research also emphasizes many minor field crops, such as oats and barley. Although a large amount of plant breeding R&D shifted from the public sector to the private sector, ample research opportunities still exist for both sectors.

Introductions and Trials of New Varieties Are Increasing Over Time

Improved plant varieties are a product of research and development. Seeds embody the scientific knowledge needed to produce a new plant variety with desirable attributes, such as higher yield potential, greater disease resistance, or improved quality. The number of plant variety protection (PVP) certificates issued by USDA provides a useful indicator of plant breeding research efforts. The number of PVP certificates issued by USDA has grown rapidly since the 1970 Plant Variety Protection Act. This growth suggests the positive effects of the Act on generating private sector incentives for plant breeding R&D. Growth in PVP certificates has been highest for soybeans and corn, which together account for more than half of all certificates issued for field crops. By the end of 2002, 2,612 certificates had been issued for varieties of U.S. origin for the four major field crops, including 1,078 for soybeans, 648 for corn, 568 for wheat, and 290 for cotton. Most PVP certificates are held by the private sector.

As indicated by the number of field releases issued by USDA's Animal and Plant Health Inspection Service (APHIS) to allow breeders to pursue field testing, plant breeding research through biotechnology is increasing. Between 1987 and June 2001, APHIS received over 7,600 applications for field releases of biotech varieties and approved 6,700. The number of applications received annually by APHIS increased from 9 in 1987 to 1,206 in 1998. The majority of applications for field releases received from private and public institutions are for testing improved varieties of major crops. By mid-2001, more than 3,300 applications had been received for corn varieties. Other field release applications included 761 for potatoes, 601 for soybeans, 532 for tomatoes, 481 for cotton, and 209 for wheat.

After undergoing years of field tests, extensive review, and determination by APHIS that unconfined release of a genetically modified organism does not pose a significant risk to agriculture or the environment, the organism in question is no longer considered a regulated article and can be moved and planted without APHIS authorization. As of mid-2001, APHIS had received 79 petitions for deregulation and had granted 53. Thirty-six percent of these deregulated varieties have herbicide-tolerance traits, 20 percent have insect-resistance traits, and 19 percent have traits to improve product quality.

INTRODUCTION

In 1798, Thomas Malthus predicted that geometric population growth and arithmetic food production increases would lead to chronic food shortages, with dire consequences for the future of humanity. Those predictions have failed to materialize largely because worldwide agricultural production has increased enough to accommodate a sixfold increase in population. In particular, the unprecedented growth in crop yields and agricultural total factor productivity (ratio of total outputs to total inputs) over the past 70 years owes much to a series of mechanical, chemical, and biological innovations driven by agricultural research and development (figure 1).

To a large extent, these yield increases resulted from a series of biological innovations embodied in seeds. The first, and possibly most significant, innovation was the development of hybrid crops, particularly corn, in the United States in the 1930s. Improved varieties also raised yields in many other crops. Developing countries also adopted high-yielding crop varieties, spurring the Green Revolution of the 1960s and 1970s. More recently, modern biotechnology, especially genetic engineering, is facilitating the development of new biological innovations embodied in seed.

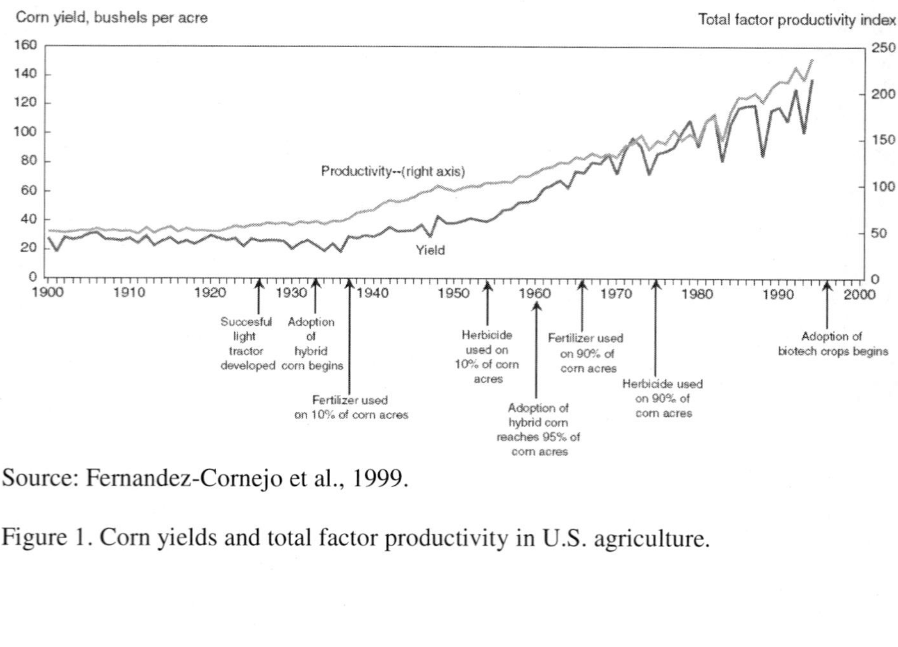

Source: Fernandez-Cornejo et al., 1999.

Figure 1. Corn yields and total factor productivity in U.S. agriculture.

BACKGROUND: THE SCIENCE OF SEED

Modern U.S. agriculture owes much to the application of science. Scientific discoveries in the fields of chemistry, engineering, and biology have contributed to the development of a highly productive and technologically innovative modern agricultural sector. Central to this modernization process has been the application of science to modern plant breeding.

The first application of modern scientific methods to plant reproduction is credited to Gregor Mendel in the mid-19th century (Sears, 1947). Up to then, farmers engaged in plant breeding in a less systematic or conscious manner, usually by exploiting chance mutations and natural selection processes. Mendel's research focused on the identification of particular traits in garden peas, and the ways in which such traits were inherited by successive generations. Mendel's work on the laws of heredity, though lost for 34 years, eventually gave rise to extensive scientific research into the inheritance of traits in other plants and crops (Jenkins, 1936, p. 483).

A significant portion of this later research focused on corn and corn hybridization. Essentially, hybridization is a traditional breeding process in which inbred lines are crossed to create seed varieties with greater yield potential than exhibited by either parent (see box on terms and concepts). Hybridization allows breeders to enhance biological characteristics more predictably and more quickly than natural selection or chance mutations. Breeders also protect their intellectual property by keeping knowledge of their hybrid varieties from being passed on to others. Early 20th century studies, including De Vries and Correns (1900) on the inherited nature of corn endosperm texture, Shull and East (1908) on hybrid vigor arising from their experiments in corn breeding, and Jones (1918) on the commercial potential of higher yielding hybrid corn, led to major breakthroughs in plant breeding (Heisey, 1999).

By the late 1920s, hybrid corn breeding programs were showing signs of success, particularly large increases in yields. Corn, as an open-pollinated crop, was well suited to the inbreeding-hybridization process. From the perspective of the farmer, hybrid corn seed had many advantages, including higher yield potential, greater uniformity in maturity, and resistance to lodging, making large-scale mechanization possible (Sprague, 1961, p. 107). From the perspective of the seed firm, hybridization had two commercial advantages. First, simple examination of a hybrid seed does not reveal its lineage, thus offering companies proprietary control over the seeds they develop. Second, the enhanced vigor of hybrid seed is not transmitted to its offspring, thereby requiring farmers to buy new seed every year to ensure continued vigor. Crops cultivated from seed saved from a hybrid crop grown in the previous year are typically less vibrant and significantly lower in yield.

The application of science to plant breeding, and specifically to corn, was a timely development in U.S. agriculture. In the early 20th century, corn was the dominant field crop in U.S. agriculture. Yet, despite the crop's importance, corn yields were stagnant at the time. Publicly supported research and development of hybrid corn seed, an effort later taken over by commercial seed companies, helped reverse this trend during the 1930s (Airy et al., 1961, p. 145). The first seed company was organized for the commercial production of hybrid corn in 1926, but hybrid corn seed production only began to expand in the early 1930s, as several new firms began production (Jenkins, 1936, p. 479). By 1960, the share of corn acreage cultivated with hybrid seed in the United States had reached 95 percent (figure 2), and almost all open-pollinated (OP) corn cultivated in the United States was replaced by hybrids by the 1960s (Fernandez-Cornejo et al., 1999; Shoemaker et al., 2001, p. 9).

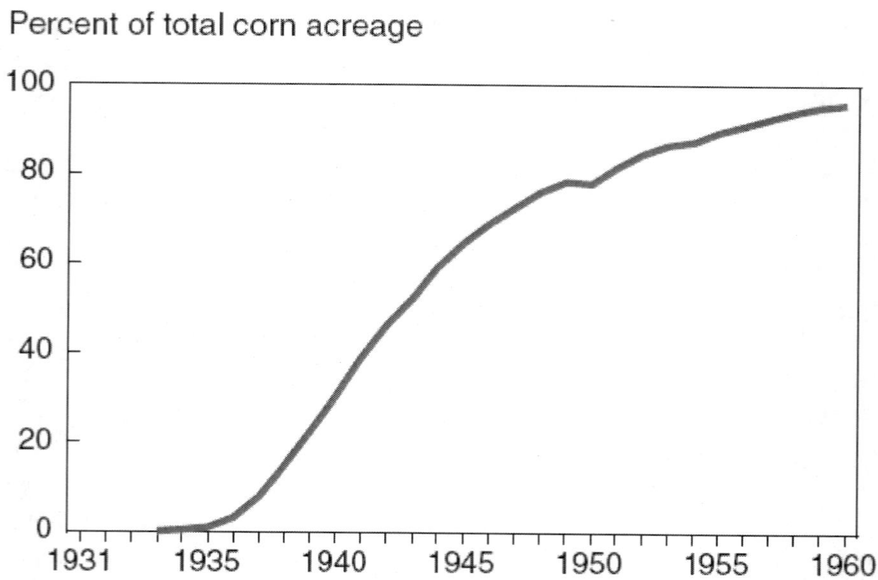

Source: *Agricultural Statistics,* NASS, USDA, various years.

Figure 2. Adoption of hybrid corn.

TERM AND CONCEPTS

Plant breeding concepts: "The essence of plant breeding is the discovery or creation of genetic variation in a plant species and the selection from within that variation of plants with desirable traits that can be inherited in a stable fashion. The plant breeders' final selections of superior plants will form the basis of one or more plant varieties.Plant breeders use all available technology both to create genetic variation and to select from within that variation.

"Different types of plant variety have been developed, depending upon the physiology of the plants of each species and the ways in which the plants of the species can be reproduced. For example, varieties of rose and potato can be reproduced vegetatively, that is to say, can be reproduced by using a part of a plant as the basis for producing another complete plant. Rose varieties can be reproduced by propagating a bud or a cutting from a plant of the variety. Potato varieties are normally reproduced by propagating a tuber of the variety.

"Varieties of grasses and most vegetables and cereals are reproduced sexually, that is by pollination of the female part of a flower (the stigma) by pollen from the male part of a flower (the anther). Here, however, one must make a distinction. The plants of some species, for example wheat, will tolerate, through successive generations, the fertilization of the stigma by pollen from the anthers of the same flower or from another flower on the same plant without loss of vigor. Plant varieties of such species can be based upon a single plant or on a small number of plants which will reproduce themselves precisely through successive generations. All the plants of a variety of this kind, known as "self-pollinated" varieties, will be genetically the same or very similar.

"The plants of many species are not adapted to self-fertilization or cannot tolerate self-fertilization through successive generations and will become less vigorous if forced to self-pollinate (they will suffer from "inbreeding depression"). In these plants, the female part of the flower must be fertilized by the pollen from another flower, or from a flower of another plant. Varieties of such species, known as "cross-pollinated" varieties, are populations of plants based upon the controlled cross-pollination of a sufficient number of selected diverse, superior plants to secure enhanced performance without suffering in-breeding depression.

"Yet a further category of variety is based upon the controlled cross-pollination of parent lines, so that the seed resulting from the cross-pollination inherits its genetic make-up from the parent lines. Such varieties, known as "hybrids," will typically exhibit greater vigor ("hybrid vigor") than the parent lines on which they are based, resulting, for example, in plants with higher yields, better resistance to stress, etc. The same controlled cross-pollination must be repeated each time the seed of those varieties is produced." (UPOV, 2003).

Agricultural biotechnology is the application of scientific techniques, including genetic engineering, to create, improve or modify plants, animals and microorganisms. Agricultural biotechnology improves upon conventional techniques, such as selective breeding, by enabling scientists to move genes and the desirable traits that they express with greater efficiency and precision (USDA, 2003).

Genotypes are the genetic traits or characteristics expressed by a particular variety (UPOV).

Germplasm is the genetic material that contains a variety's inherited characteristics.

Cross-pollinating species are those in which the pollen is primarily dispersed from one plant to another.

Self-pollinating species are those in which the pollen is primarily dispersed within the same plant.

Organisms are living things that may be categorized for purposes here as plants, animals, or microorganisms, including bacteria. An organism is categorized based on such factors as its structure, mobility, source of nutrition, or cell structure (UPOV; USDA 2000).

Sexual reproduction includes any production of a variety by seed but does not include the production of a variety by tuber propagation (PVP Act, Chapter 4, Sec. 41, USDA, Agricultural Marketing Service (AMS), 2001).

Tuber propagation means propagated by a tuber or a part of a tuber (PVP Act, Chapter 4, Sec. 41, USDA, AMS, 2001).

Variety refers to a plant grouping within a single botanical taxon [taxonomic group] of the lowest known rank, that, without regard to whether the conditions for plant variety protection are fully met, can be defined by the expression of the characteristics resulting from a given genotype or combination of genotypes, distinguished from any other plant grouping by the expression of at least one characteristic and considered as a unit with regard to the suitability of the plant grouping for being propagated unchanged. A variety may be represented by seed, transplants, plants, tubers, tissue culture plantlets, and other matter (PVP Act, Chapter 4, Sec. 41, USDA, AMS, 2001).

The commercial success of hybrid corn in the United States was followed by such plant breeding advances as hybrid sorghum and improved varieties of soybeans and cotton. The first commercial seed field of hybrid sorghum was planted in 1955; by 1960, 70 percent of the U.S. sorghum acreage was planted with hybrid seed (Airy et al., 1961, p. 145). Sorghum is now mostly grown from hybrid seed. Other vegetables, including onions, spinach, tomatoes, and cabbage, are also grown from hybrid seed (Emsweller, 1961).

The application of science to plant breeding also produced significant gains in crop yields in other countries. During the late 1960s and early 1970s, scientific breakthroughs in the breeding of key crops, such as rice and wheat, boosted production in many developing countries. When combined with the proper use of fertilizers, pesticides, and other inputs, these new varieties offered yield increases of 1 percent per year through the 1990s (Morris, 1998, pp. 3-4). The unprecedented growth in agricultural output in the developing world is often referred to as the "Green Revolution" and has played an important part in improving food security in such countries as India and China.

Despite its major role in plant breeding, the development of hybrids through interbreeding can require up to 12 years to develop market-ready seeds. Even then, those hybrids may still generate only limited desired traits or, possibly, unwanted characteristics (Gould, 1983; Ollinger and Fernandez-Cornejo, 1995, p. 17). Scientific discoveries in the field of genetics, beginning with Watson and Crick's postulate on the double helix model for DNA in 1953 and continuing through the creation of the first genetically engineered (GE) plant in 1982 (Shoemaker et al., 2001, p. 9), have significantly reduced the number of residual unwanted characteristics that often result from traditional plant breeding crosses, thus increasing the speed at which breeders can develop desirable new varieties.

GE crops are classified into one of three generations (Panos, 1998): crops with enhanced input traits, such as herbicide tolerance, insect resistance, and resistance to environmental stresses, such as drought; crops with added-value output traits, such as nutrient-enhanced seeds for feed; and crops that produce pharmaceuticals, bio-based fuels, and products beyond traditional food and fiber. At present, the adoption of GE crops is generally limited to those with first-generation traits. The most common herbicide-tolerant (HT) crops are characterized by resistance to the herbicide glyphosate. Commercially available HT crops include soybeans, corn, canola, and cotton and became available to a limited extent in 1996. The share of HT soybeans to total U.S. soybean acreage grew from 17 percent in 1997 to 68 percent in 2001 and 81 percent in 2003. HT cotton expanded from 10 percent of total U.S. cotton acreage in 1997 to 56 percent of cotton acreage in 2001 and 59 percent in 2003 (figure 3). Insect-resistant GE crops in use today incorporate a gene from the soil bacterium *Bacillus thuringiensis* (Bt). The bacterium produces a protein toxic to lepidopteran insects. Plants produce the toxic protein throughout their life, providing them with long-term protection. The Bt gene has been incorporated in corn to protect the crop against the European corn borer, and in cotton to protect against the tobacco budworm, bollworm, and pink bollworm (Fernandez-Cornejo and McBride, 2000). Bt corn increased from 1 percent of total U.S. corn acreage in 1996 to 26 percent in 1999, fell to 19 percent in 2000 and 2001, and climbed back to 29 percent in 2002. Bt cotton increased from 15 percent of cotton acres in 1996 to 37 percent in 2001 and 41 percent in 2003.

Improved Seed Is a Major Contributor to Crop Yield Gains and Agricultural Productivity

Over the past 70 years, there has been a remarkable increase in the yields of all major field crops in the United States, and more than half of the yield gains are attributed to genetic improvements achieved by plant breeders.

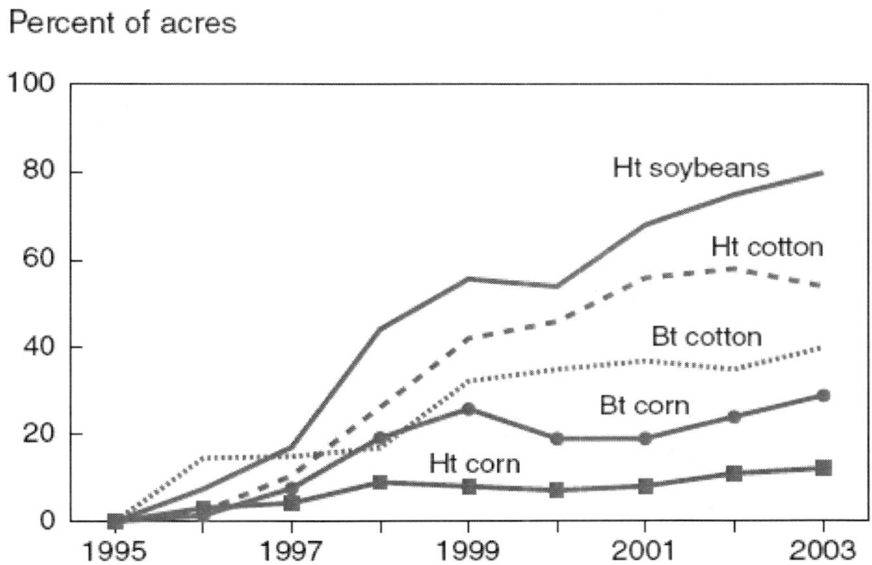

Source: Fernandez-Cornejo, 2003.

Figure 3. Adoption of biotech crops in the United States.

Among the four major field crops examined in this report, yield increases have been most significant for corn; the next highest increases were for cotton, soybeans, and wheat (figure 4). Average per acre yields for corn in the United States rose from 20 bushels in 1930 to about 70 bushels in 1970 and reached 140 bushels by the mid-1990s (figure 5). Soybean yields have also increased substantially since the 1930s (figure 6). Overall per acre yields for soybeans increased from about 13 bushels in 1930 to nearly 40 bushels in the mid-1990s.

Cotton yields stagnated from 1866 to 1935. Yields grew rapidly from 1935 to 1960 as higher yielding cultivars were introduced together with synthetic fertilizers and pesticides (figure 7) (Meredith and Bridge, 1984). After reaching a plateau during the 1960s and early 1970s, cotton yields resumed their rapid growth. Overall, cotton yields rose nearly fourfold during 1930-98 (figures. 4, 7). Wheat breeding has relied heavily on genetic improvement of self-pollinating varieties. Though wheat yields have grown steadily since 1950, the rate of growth has not been as high as that of corn and cotton (figure 8). Still, overall wheat yields increased 2.5-fold during the period.

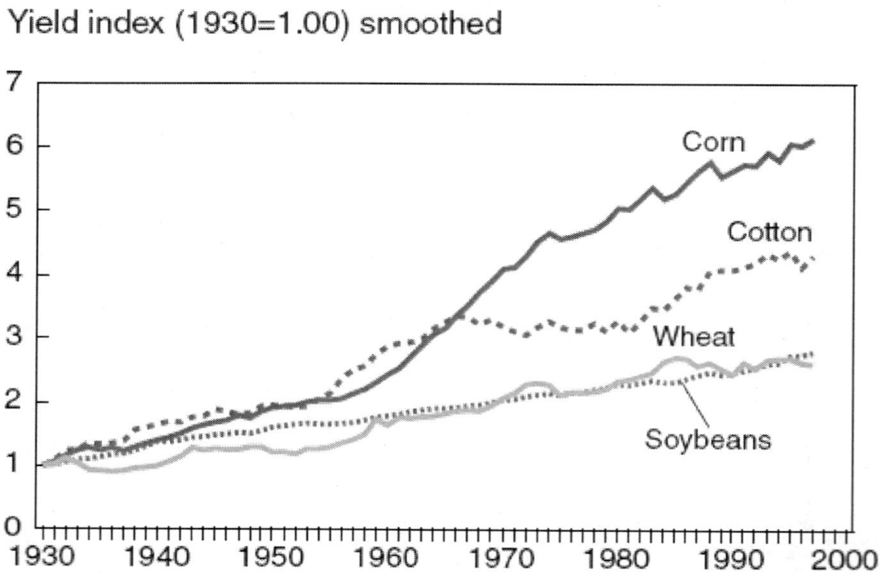

Source: *Agricultural Statistics,* NASS, USDA, various years.

Figure 4. Yields for major crops.

Contribution of Plant Breeding

Crop yields have benefited from genetic improvements through plant breeding as well as from improved pest management, mechanization, and fertilizer use. Still, extensive evidence suggests that crop yields have benefited the most from plant breeding, which includes the use of improved hybrids and varieties. Studies of the determinants of increased crop yields for corn, soybeans, and wheat conclude that 50 percent or more of the overall yield gain for each crop can be attributed to genetic improvements of plant varieties. For the same crops, the annual rate of yield gain due to plant breeding improvements is 1-3 percent per year (Duvick, 1992, p. 291).

Different estimates of plant breeding's contribution to crop yield gains reflect differences in the period covered and the complex nature of the research. Thirtle (1985, in Fuglie et al., 1996, p. 44) concludes that between 1939 and 1978, biological inputs (improved seed varieties and changes in agronomic practices) increased average annual yields by 1.7 percent for corn,

1.1 percent for soybeans, 0.5 percent for cotton, and 1.5 percent for wheat. Over the entire period, biological inputs contributed to 50 percent of the yield growth in corn, 85 percent for soybeans, 24 percent for cotton, and 75 percent for wheat. Fehr (1984, in Fuglie et al., 1996, p. 44) estimates that genetic improvements accounted for 89 percent of the gain in corn yields between 1930 and 1980, a 90-percent gain in soybean yields between 1902 and 1997, 67 percent of the gain for cotton yields between 1936 and 1960, and a 50-percent gain for wheat yields between 1958 and 1980.

The effects of other contributing factors to crop yield gains—fertilizers, pesticides, machinery, and labor— vary widely. For example, Cardwell (1982, reported in Duvick, 1992, p. 29) concludes that, since 1930, better weed control has accounted for 23 percent of corn yield gains in Minnesota, synthetic nitrogen fertilizer has accounted for 19 percent, and plant breeding has accounted for 59 percent (16 percent for the shift from open-pollinated to hybrid seeds, and 43 percent for other breeding improvements).

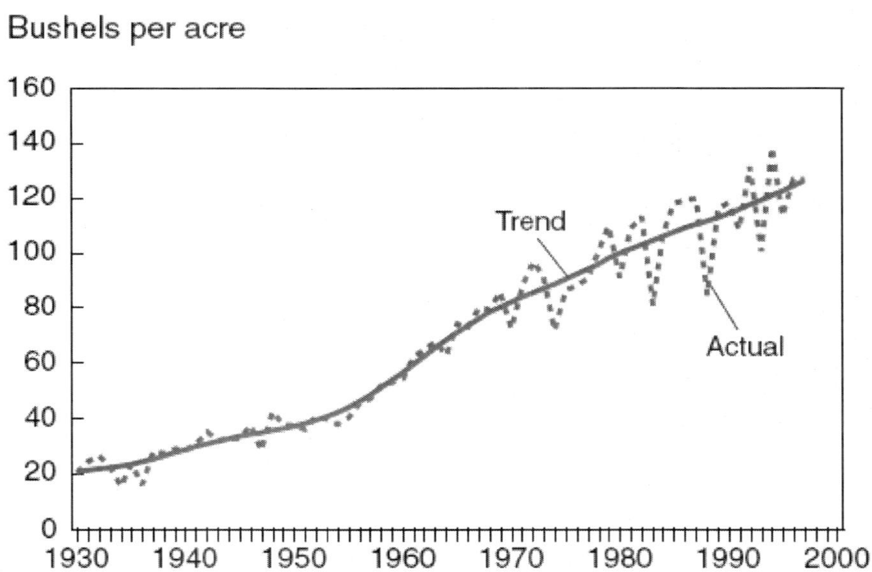

Source: *Agricultural Statistics,* NASS, USDA, various years.

Figure 5. Corn yields.

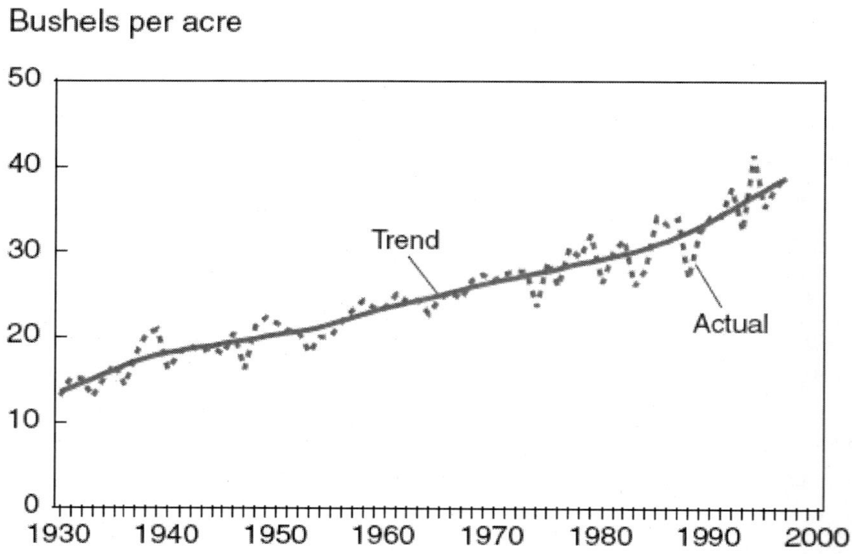

Source: *Agricultural Statistics*, NASS, USDA, various years.

Figure 6. Soyebean yields.

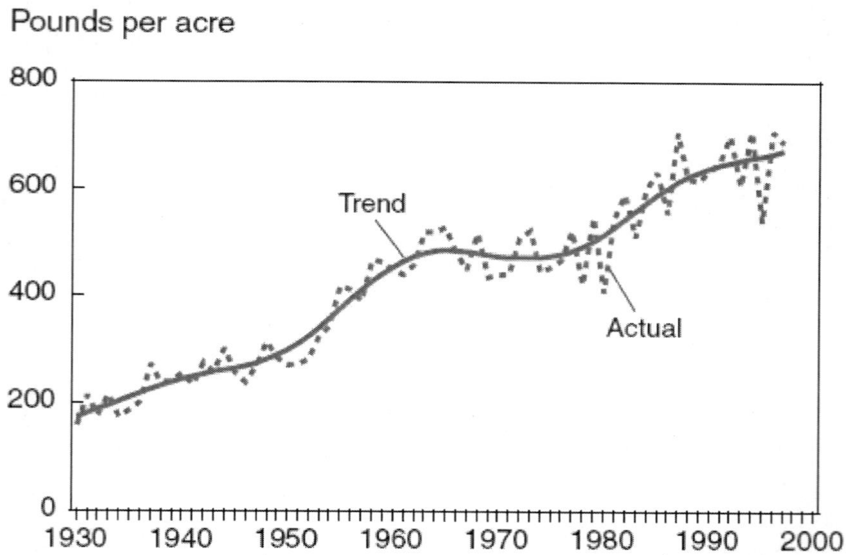

Source: *Agricultural Statistics,* NASS, USDA, various years.

Figure 7. Cotton yields.

The remarkable yield gains over the past 50 years are observed not only in crops cultivated under ideal conditions, where complementary inputs, such as soil fertility and water supply, are optimal, but also in crops cultivated under less ideal conditions, where yield-improving inputs are not optimized due to such conditions as drought or pests. Many new varieties possess genetic qualities that improve seed productivity in both favorable and unfavorable environments. For example, the improved root strength in newer hybrid corn varieties increased the plant's ability to resist stalk-rot fungi, heat, drought, limited nitrogen nutrition, and pests, such as the European corn borer (Duvick, 1992, pp. 292-293). Still, despite the improved plant varieties, crop yields have been subject to periods of stagnation or slow growth. Cotton yields, for example, stagnated between 1960 and 1980 before increasing again in the 1980s (Fuglie et al., 1996, p. 44).

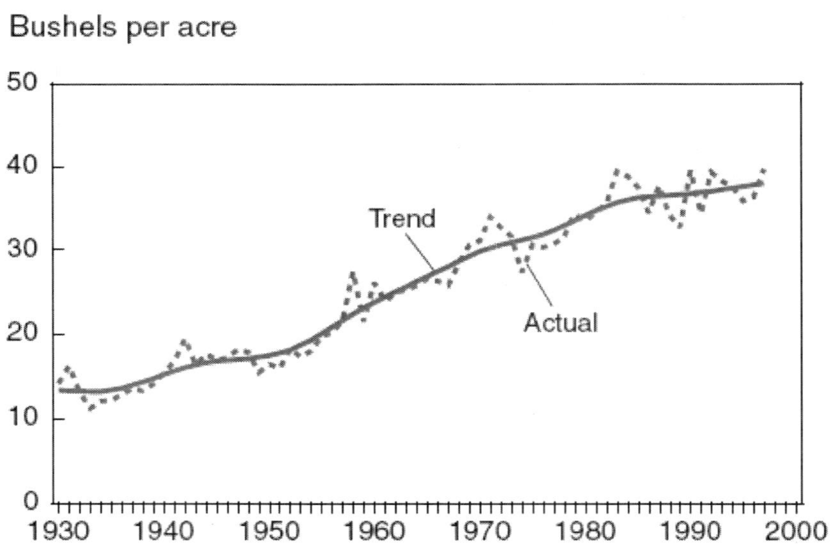

Source: *Agricultural Statistics,* NASS, USDA, various years.

Figure 8. Wheat yields.

Purchased Seed Use, Purchase Value, and Trade Have Grown in Recent Decades

Demand for seed is primarily determined by the profitability of the crop that the seed produces. Seed use by U.S. farmers depends on several factors, including the acreage under cultivation, the seeding rate per cultivated acre, cropping practices, and variations in geographic location and agroclimatic conditions.

Seed Market Size and Value

The U.S. seed market is rapidly growing in size and value. U.S. farmers used over 6.5 million tons of seed for major field crops in the 1996/97 crop marketing year (table 1). In 1997, the estimated total value of the commercial U.S. seed market was $5.7 billion, roughly 20 percent of the world market in seed (table 2) (FIS/ASSINSEL, 2001). The U.S. seed market is the largest seed market worldwide. China, at $3.0 billion, and Japan, at $2.5 billion, are the next largest seed markets.

Total seed expenditures by U.S. farmers rose from about $500 million in 1960 to over $6.7 billion in 1997 (table 3, figure 9). In real terms, seed expenditures climbed about 2.5-fold in the same period, despite minimal real increases in the index of seed prices paid by farmers (about 7 percent in 38 years, table 3). Similarly, when measured as a share of total farm expenditures, seed expenditures increased from 2 percent in 1970 to 4 percent in 1997 (figure 10).

A large portion of the increase in real seed expenditures may be explained by increases in the share of seed purchased by U.S. farmers from seed, particularly for major field crops, which account for the largest share of seed purchased (tables 4, 5). In 1997, 81 percent of all U.S. soybean acreage (up from 55 percent in 1982) and 78 percent of all U.S. cotton acreage (up from 50 percent in 1982) were cultivated with purchased seed. These increases, in turn, can be explained by increases in seed productivity attributable to scientific improvements in plant breeding. Seeding rates have also increased (tables 6, 7).

Seeding Rates and Costs for Major Field Crops

Seeding rates, acreage used in crop production, and share of seed purchased are the key agricultural factors affecting the demand for purchased seed in U.S. agriculture. Seeding rates, seed prices, and seed costs per acre for corn, soybeans, cotton, and wheat over time are presented in tables 6 and 7. For 1972-95, the average seed cost per acre is derived by multiplying seeding rate per acre and seed price. For 1986-97, average seeding rate and seed cost per acre are based on data from the Cropping Practices Surveys and the Agricultural Resource Management Survey (ARMS).

Table 1. U.S. seed use for major field crops, by crop and marketing year Seed use

Crops	Seed use										
	1986/87	1987/88	1988/89	1989/90	1990/91	1991/92	1992/93	1993/94	1994/95	1995/96	1996/97
	Thousand tons										
Corn	468	482	523	529	540	566	528	563	510	571	580
Sorghum	44	36	42	36	34	48	39	34	34	45	39
Soybeans	1,653	1,684	1,766	1,664	1,701	1,705	1,726	1,914	1,877	1,929	2,064
Barley	429	376	360	324	350	310	314	283	266	283	278
Oats	608	505	433	374	306	290	285	240	205	194	219
Wheat	2,520	2,550	3,090	3,009	2,787	2,931	2,973	2,889	2,679	3,123	3,075
Rice	130	150	150	180	180	195	180	210	205	186	185
Cotton, upland	93	106	89	94	110	120	115	117	118	144	125
Total	5,945	5,889	6,453	6,210	6,013	6,165	6,160	6,250	5,894	6,475	6,566

Sources: ERS estimates, based on data from the Agricultural Resource Management Survey and the Cropping Practices Surveys(USDA, ERS, 2002).

Table 2. Estimated values of commercial markets for seed

Country	Internal commercial market[1]
	Million dollars
UnitedStates	5,700
China	3,000
Japan	2,500
Comonwealth of Independent States[2]	2,000
France	1,370
Brazil	1,200
Germany	1,000
Argentina	930
India	900
Italy	650
United Kingdom	570
Canada	550
Poland	400
Mexico	350
Spain	300
Netherlands	300
Australia	280
Hungary	200
Denmark	200
Sweden	200
Other	1,967
Total	24,567

[1] Data provided in this table are the most recent figures (available at the Secretariat of the International Seed Federation). Data refer to different years depending on the countries. Data include planting materials. The total represents the sum of the commercial seed markets of the countries listed by FIS/ASSINSEL. See source for details. The commercial world seed market is assessed by FIS/ASSINSEL at approximately US$30 billion.

[2] States from the former Soviet Union (Armenia, Azerbaijan, Belarus, Georgia, Kazakhstan, Kyrgyzstan, Moldova, Russia, Tajikistan, Turkmenistan, Ukraine, and Uzbekistan).

Source: FIS/ASSINSEL (2000).

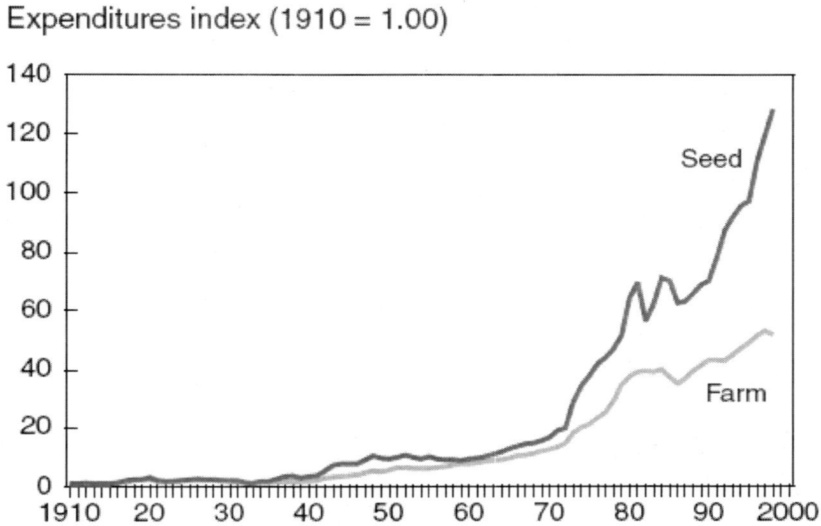

Source: Data sources provided in table 3.

Figure 9. Farm and seed expenditure indices.

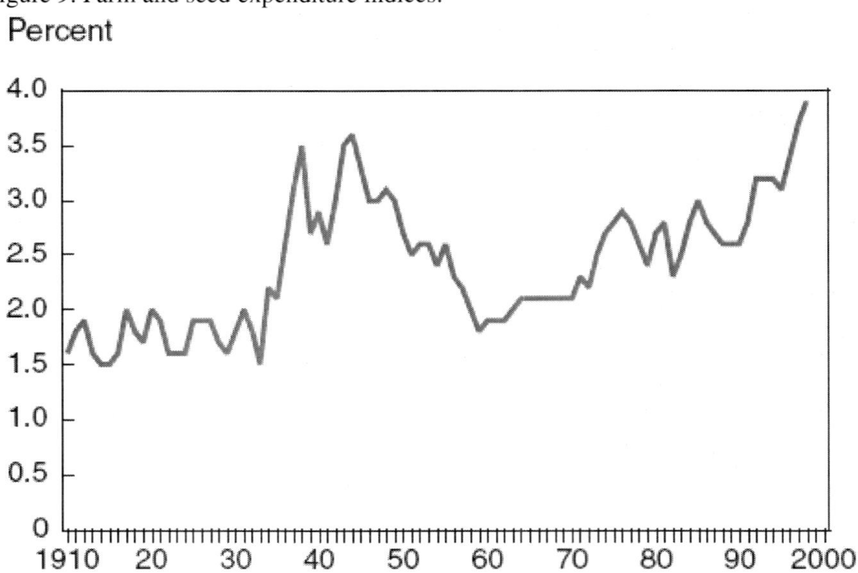

Source: Data sources provided in table 3.

Figure 10. Seed expenditures' share of total farm expenditures.

Seeding rates per acre can vary, depending on geographic location and cropping practices. Seeding rates tend to change slowly over time but the acreage planted with a particular crop can vary from year to year. For example, in 1996-97, planted acreage for soybeans, corn, and oats increased while acreage for cotton, sorghum, and wheat decreased. Demand for purchased seed may also be affected by decreases in the use of saved seed for such crops as soybeans. As a result, seed use for each of the crops varies from year to year (table 1). But, total seed consumption for the major field crops remained fairly constant between the 1986-87 and 1996-97 marketing years, ranging from 5.9 to 6.6 million tons per year.

Corn. Corn seeding rates and costs vary substantially by State. States where crops are mostly irrigated, such as California, or receive abundant rainfall, such as in the eastern Corn Belt, support heavier seeding rates, and consequently, higher seed costs per acre than other States. Within the Corn Belt, average seeding rates and seed costs per acre in selected States vary considerably, primarily because of differences in soil productivity, moisture availability, and seed price (table 6). For example, in 1989, Minnesota had the highest seeding rate (25,800 kernels) per acre. As a result, Minnesota had the highest average cost per acre at $21.44. South Dakota had typically lower and more variable precipitation than other corn-growing States, resulting in lower seeding rates.

The price of seed per unit also affects seeding costs. For example, in 1988 and 1989, the average corn seeding rate was constant at 24,100 kernels per acre, but the average seeding cost per acre for the 10 leading corn-producing States rose from $18.64 in 1988 to $20.40 in 1989, due to an 11-percent increase (nominal; 5 percent in real terms) in corn seed price.

Soybeans. Soybean seeding rates per acre showed little year-to-year variation. For example, the seeding rate per acre was constant at 57 pounds from 1980 to 1985; the rate increased to 59 pounds in 1986 and 60 pounds in 1989 (table 6). Seeding costs per acre, however, varied significantly due to changes in the average seed price of soybeans. Seeding rates per acre were the same (70 pounds) in 1995 and 1996, but seeding costs per acre were higher in 1996 ($18.62) than in 1995 ($16.80) due to higher seed prices. Soybean seeding rates tend to be lower in Southern States, such as Arkansas, Georgia, Mississippi, Tennessee, and Louisiana. Northern States, on the other hand, have higher seeding rates and yields, and therefore have higher seed costs per acre than Southern States.

Table 3. U.S. farm seed expenditures and farm seed price index

Year	Expenditures on seed	Total farm production expenditures	Share of total farm production expenditures	Farm seed price index 1984=100	Prices paid by farmers, index 1984=1.00	Real farm seed price index (relative to prices paid by farmers)	Real expenditures on seed
	——Million dollars—		*Percent*				*Million 1989 dollars*
1960	519	27,376	1.9	0.2247	0.2657	0.8456	1,953
1961	545	28,590	1.9	0.2226	0.2675	0.8321	2,037
1962	565	30,279	1.9	0.2311	0.2719	0.8499	2,078
1963	619	31,598	2.0	0.2460	0.2764	0.8902	2,240
1964	661	31,812	2.1	0.2439	0.2772	0.8797	2,384
1965	720	33,650	2.1	0.2524	0.2852	0.8850	2,524
1966	760	36,508	2.1	0.2492	0.2967	0.8398	2,561
1967	814	38,181	2.1	0.2535	0.3020	0.8392	2,695
1968	831	39,525	2.1	0.2630	0.3091	0.8509	2,688
1969	871	42,115	2.1	0.2684	0.3242	0.8278	2,687
1970	928	44,452	2.1	0.2843	0.3384	0.8404	2,743
1971	1,072	47,107	2.3	0.3131	0.3543	0.8837	3,026
1972	1,115	51,688	2.2	0.3419	0.3764	0.9081	2,962
1973	1,617	64,554	2.5	0.4228	0.4349	0.9722	3,718
1974	1,941	70,980	2.7	0.5453	0.4942	1.1032	3,927
1975	2,138	75,043	2.8	0.6198	0.5430	1.1415	3,938

Table 3. (Continued)

Year	Expenditures on seed	Total farm production expenditures	Share of total farm production expenditures	Farm seed price index 1984=100	Prices paid by farmers, index 1984=1.00	Real farm seed price index (relative to prices paid by farmers)	Real expenditures on seed
	——Million dollars—		*Percent*				*Million 1989 dollars*
1976	2,366	82,742	2.9	0.6113	0.5784	1.0569	4,091
1977	2,484	88,884	2.8	0.6613	0.6103	1.0837	4,070
1978	2,638	103,250	2.6	0.6922	0.6599	1.0490	3,998
1979	2,904	123,304	2.4	0.7252	0.7511	0.9656	3,866
1980	3,220	133,139	2.7	0.7838	0.8397	0.9335	3,835
1981	3,428	139,444	2.8	0.9137	0.9167	0.9967	3,739
1982	3,172	140,306	2.3	0.9340	0.9548	0.9782	3,322
1983	2,987	139,608	2.5	0.9350	0.9779	0.9562	3,055
1984	3,548	141,876	2.8	1.0000	1.0000	1.0000	3,548
1985	3,941	132,433	3.0	1.0106	0.9894	1.0215	3,983
1986	3,511	125,084	2.8	0.9776	0.9681	1.0098	3,627
1987	3,259	130,992	2.7	0.9819	0.9832	0.9987	3,315
1988	4,060	139,907	2.6	0.9915	1.0549	0.9399	3,849
1989	4,397	146,662	2.6	1.0937	1.1169	0.9792	3,937
1990	4,518	153,302	2.6	1.0777	1.1603	0.9288	3,894
1991	5,113	153,279	2.8	1.0469	1.1816	0.8860	4,327
1992	4,913	152,940	3.2	1.0394	1.1940	0.8705	4,115

Year	Expenditures on seed	Total farm production expenditures	Share of total farm production expenditures	Farm seed price index 1984=100	Prices paid by farmers, index 1984=1.00	Real farm seed price index (relative to prices paid by farmers)	Real expenditures on seed
	——Million dollars—		*Percent*				*Million 1989 dollars*
1993	5,163	160,506	3.2	1.0671	1.2232	0.8724	4,221
1994	5,373	167,504	3.2	1.1406	1.2542	0.9094	4,284
1995	5,462	174,120	3.1	1.1587	1.2879	0.8997	4,241
1996	6,212	182,439	3.0	1.2162	1.3561	0.8968	4,581
1997	6,711	188,443	3.7	1.2556	1.3942	0.9006	4,814

Sources: Expenditures for 1910-80 from Lucier et al. (1986); expenditures for 1981-97 from *Agricultural Statistics* (USDA, NASS, various years). Price indices from *Agricultural Statistics* (USDA, NASS, various years).

Table 4. Estimated U.S. market and purchased seed value, 1982

Seed stock	Total market value of seed[1]	Share of total market value of seed	Share of acres planted with purchased seed	Market value of purchased seed	Share of market value of purchased seed
	Million dollars	*Percent*	*Percent*	*Million dollars*	*Percent*
Corn	1,294	27.8	95.0	1,230	40.2
Soybeans	954	20.5	55.0	525	17.2
Cotton	144	3.1	50.0	72	2.4
Wheat	888	19.1	10.0	89	2.9
Total, major crops	3,280	70.6	58.4	1,915	62.6
Other crops	1,369	29.4	83.4	1,142	37.4
Total	4,649	100.0	65.8	3,057	100.0

[1] Including the market value of saved seed.

Source: McMullen (1987b, pp.86-87).

Table 5. Estimated U.S. market and purchased seed values, major crops, 1997

Seed stock	Acres planted	Seeding rate per acre[1]	Seed cost per acre[2]	Total market value of seed[3]	Share of total market value of seed	Share of acres planted with purchased seed	Market value of purchased seed	Share of market value of purchased seed
	Million acres	*Kernels or lbs*	*Dollars*	*Million dollars*	*Percent*	*Percent*	*Million dollars*	*Percent*
Corn	80.2	27,665	28.87	2,316	35.1	100.0	2,316	40.2
Soybeans	70.9	70	18.78	1,331	20.2	81.0	1,078	18.7
Cotton	13.3	14	10.49	139	2.1	78.0	109	1.9
Wheat[4]	71.0	83	12.21	867	13.1	37.0	321	5.6
Total, major crops	235.3			4,653	70.6		3,823	66.3

[1] For corn, kernels per acre was used; for soybeans, wheat, and cotton, pounds per acre was used. Based on figures from select States, from USDA, Agriculture Resource Management Survey (ARMS), 1998.

[2] Seed cost per acre is based on USDA's ARMS data.

[3] Including the market value of saved seed.

[4] Compiled from statistics provided in this report on winter wheat, spring wheat, and durum wheat. Seed in grate, seed cost per acre, and share of acres planted with purchased seed are based on the weighted averages of the three types of wheat. For durum, the seed cost per acre was not available and was estimated according to the percent change in the unit price of the crop.

Source: Acres planted from *Agricultural Statistics* (USDA,NASS, 1998); other data from ARMS.

Table 6. Seeding rate, cost per acre, and share of acres with purchased seed, major field crops in surveyed States

Y ear	Corn					Soybeans				Cotton				Real price of seed		
	Average rate per acre	Average cost per acre	Acres with Purchased seed	Seed price Per bushel	Average rate per acre	Average cost per acre	Acres with Purchased seed	Seed price Per bushel	Average rate per acre	Average cost per acre	Acres with purchased seed	Seed price per bushel	Deflator: (Index of prices paid by U.S. farmers) 1984=1.00	Corn per bushel	Soybean per bushel p	Cotton er 100 lbs
	Kernels	*Dollars*	*Percent*	*Dollars*	*Pounds*	*Dollars*	*Percent*	*Dollars*	*Pounds*	*Dollars*	*Percent*	*Dollars*		*1984 dollars*		
1972	20,955	4.90	NA	21.50	68	5.90	NA	5.21	25.00	4.35	NA	17.40	0.38	57.11	13.84	46.22
1973	20,955	5.03	NA	22.20	68	9.89	NA	8.75	25.00	4.56	NA	18.30	0.43	51.05	20.12	42.08
1974	20,955	5.66	NA	25.00	68	10.74	NA	9.50	25.00	6.33	NA	25.30	0.49	50.58	19.22	51.19
1975	22,110	8.74	NA	36.50	63	11.03	NA	10.50	26.00	8.74	NA	33.60	0.54	67.22	19.34	61.88
1976	22,110	8.74	NA	36.50	63	7.77	NA	7.40	26.00	7.41	NA	28.50	0.58	63.11	12.79	49.27
1977	22,110	9.57	NA	40.00	63	13.65	NA	13.00	26.00	7.88	NA	30.30	0.61	65.54	21.30	49.65
1978	22,110	10.29	NA	43.00	63	12.10	NA	11.50	26.00	8.06	NA	31.00	0.66	65.16	17.43	46.98
1979	22,110	10.88	NA	45.50	63	12.60	NA	12.00	26.00	8.61	NA	33.10	0.75	60.58	15.98	44.07
1980	21,945	12.46	NA	52.50	57	9.88	NA	10.40	26.00	9.15	NA	35.20	0.84	62.52	12.39	41.92
1981	21,945	14.23	NA	60.00	57	13.30	NA	14.00	24.00	9.74	NA	40.60	0.92	65.45	15.27	44.29
1982	21,879	15.12	NA	63.70	57	10.17	NA	10.70	24.00	9.98	NA	41.60	0.95	66.71	11.21	43.57
1983	21,896	15.26	NA	64.60	57	9.60	NA	10.10	24.00	10.27	NA	42.80	0.98	66.06	10.33	43.77
1984	21,900	16.67	NA	70.20	57	12.73	NA	13.40	24.00	11.52	NA	48.00	1.00	70.20	13.40	48.00
1985	21,912	16.96	NA	71.80	57	13.31	NA	11.90	24.00	11.57	NA	48.20	0.99	72.57	12.03	48.72
1986	23,800	19.09	100	65.60	59	10.44	73	10.80	18.13	7.81	68	46.60	0.97	67.76	11.16	48.13
1987	24,000	18.30	100	64.90	59	10.05	73	11.30	19.09	8.46	81	48.10	0.98	66.01	11.49	48.92
1988	24,100	18.64	100	64.20	62	12.86	73	11.90	18.00	8.36	86	47.70	1.05	60.86	11.28	45.22
1989	24,100	20.40	100	71.40	60	15.52	68	14.70	18.00	8.17	67	50.10	1.12	63.93	13.16	44.86
1990	24,700	20.50	100	69.90	62	14.20	71	12.50	17.00	7.80	70	54.30	1.16	60.24	10.77	46.80
1991	24,906	20.79	100	70.20	64	15.07	73	12.80	17.00	8.11	66	58.20	1.18	59.41	10.83	49.26

Y ear	Corn					Soybeans				Cotton			Deflator: (Index of prices paid by U.S. farmers) 1984=1.00	Real price of seed		
	Average rate per acre	Average cost per acre	Acres with Purchased seed	Seed price Per bushel	Average rate per acre	Average cost per acre	Acres with Purchased seed	Seed price Per bushel	Average rate per acre	Average cost per acre	Acres with purchased seed	Seed price per bushel		Corn per bushel	Soybean per bushel p	Cotton er 100 lbs
	Kernels	*Dollars*	*Percent*	*Dollars*	*Pounds*	*Dollars*	*Percent*	*Dollars*	*Pounds*	*Dollars*	*Percent*	*Dollars*		*1984 dollars*		
1992	25,304	21.35	100	71.80	65	15.40	73	12.40	16.00	8.74	74	59.70	1.19	60.14	10.39	50.00
1993	25,564	22.72	100	72.70	NA	16.29	73	12.40	16.00	9.39	69	62.70	1.22	59.43	10.14	51.26
1994*	25,824	23.54	100	73.40	NA	17.25	72	13.60	16.00	9.49	66	63.50	1.25	58.52	10.84	50.63
1995*	26,588	24.50	100	77.10	70	16.80	71	13.40	15.00	9.67	64	68.20	1.29	59.87	10.40	52.96
1996**	27,500	25.97	100	77.70	70	18.62	76	14.80	14.00	9.85	66	73.00	1.36	57.30	10.91	53.83
1997**	27,665	NA	100	83.50	70*	NA	81	16.10	14.00	NA	78	74.90	1.39	60.07	11.58	53.88

* States included: corn: IL, IN, IA, MI, MN, MO, NE, OH, SD, and WI; soybeans: IL, IN, IA, MN, MO, NE, OH, AR, GA, KY, LA, MS, NC, and TN; cotton: AR, GA, CA, LA, MS, and TX.

**States included: corn: IL, IN, IA, MI, MN, MO, NE, OH, SD, and WI; soybeans: IL, IN, IA, MN, MO, NE, OH, AR, LA, MS, and TN; cotton: AR, CA, GA, LA, MS, TN, and TX.

NA = not available

Sources: Average rate per acre is based on data from Cropping Practices Surveys and Agricultural Resource Management Survey (ARMS). Cost per acre for 1972-85 is computed by multiplying rate per acre (pounds) with the price of seed per pound; cost per acre for 1987-96 is based on data from Cropping Practices Surveys and ARMS. Seed prices and deflator are from *Agricultural Statistics* (USDA, NASS, various years).

Table 7. Seeding rate, cost per acre, and share of acres with purchased wheat seed in surveyed States

Year	Winter wheat				Spring wheat				Durum wheat			Deflator: (Index of prices paid by U.S. farmers) 1984=1.00	Real price of seed	
	Average rate per acre	Average cost per acre	Acres with Purchased seed	Price per bushel	Average rate per acre	Average cost per acre	Acres with Purchased seed	Price per bushel	Average rate per acre	Average cost per acre	Acres with Purchased seed		Winter wheat, per bushel	Spring wheat, per bushel
	Pounds	*Dollars*	*Percent*	*Dollars*	*Pounds*	*Dollars*	*Percent*	*Dollars*	*Pounds*	*Dollars*	*Percent*		*- 1984 dollars—*	
1972	68	3.36	NA	2.97	73	3.21	NA	2.63	NA	NA	NA	0.38	7.89	6.99
1973	66	7.92	NA	7.20	74	3.97	NA	3.20	NA	NA	NA	0.43	16.56	7.36
1974	68	8.55	NA	7.50	75	10.5	NA	8.40	NA	NA	NA	0.49	15.17	17.00
1975	74	7.50	NA	6.10	75	10.4	NA	7.70	NA	NA	NA	0.54	11.23	14.18
1976	74	7.38	NA	6.00	81	8.78	NA	6.50	NA	NA	NA	0.58	10.37	11.24
1977	70	5.15	NA	4.40	86	7.44	NA	5.20	NA	NA	NA	0.61	7.21	8.52
1978	69	6.79	NA	5.90	81	6.89	NA	5.10	NA	NA	NA	0.66	8.94	7.73
1979	67	8.21	NA	7.40	72	6.48	NA	5.40	NA	NA	NA	0.75	9.85	7.19
1980	69	8.72	NA	7.40	79	8.71	NA	6.60	NA	NA	NA	0.84	8.81	7.86
1981	73	8.89	NA	7.41	79	0.53	NA	7.22	NA	NA	NA	0.92	8.08	7.88
1982	73	9.10	NA	7.42	79	9.10	NA	6.89	NA	NA	NA	0.95	7.77	7.22
1983	72	8.74	NA	7.28	87	9.70	NA	6.69	NA	NA	NA	0.98	7.44	6.84
1984	73	8.51	NA	7.40	69	7.33	NA	6.37	NA	NA	NA	1.00	7.40	6.37
1985	72	8.23	NA	7.16	79	8.03	NA	6.10	NA	NA	NA	0.99	7.24	6.17
1986	70	7.22	43	NA	89	7.67	52	5.94	93	7.13	33	0.97	NA	6.14
1987	73	6.20	40	NA	88	6.97	49	5.56	102	7.62	44	0.98	NA	5.66
1988	75	7.67	53	7.55	90	8.58	46	5.89	99	8.05	47	1.05	7.16	5.58
1989	77	9.59	41	6.57	89	8.82	40	6.71	99	10.13	47	1.12	5.88	6.01
1990	74	8.61	39	8.01	88	8.40	39	6.05	97	7.50	27	1.16	6.90	5.21
1991	74	8.65	36	6.89	89	6.52	32	4.72	100	6.66	27	1.18	5.83	3.99

1992	74	8.65	36	7.41	91	8.39	41	6.06	96	7.56	36	1.19	6.21	5.08
Year	Winter wheat				Spring wheat				Durum wheat			Deflator: (Index of prices paid by U.S. farmers) 1984=1.00	Real price of seed	
	Average rate per acre	Average cost per acre	Acres with Purchased seed	Price per bushel	Average rate per acre	Average cost per acre	Acres with Purchased seed	Price per bushel	Average rate per acre	Average cost per acre	Acres with Purchased seed		Winter wheat, per bushel	Spring wheat, per bushel
	Pounds	*Dollars*	*Percent*	*Dollars*	*Pounds*	*Dollars*	*Percent*	*Dollars*	*Pounds*	*Dollars*	*Percent*		*- 1984 dollars—*	
1993	72	8.25	40	7.73	93	8.54	40	5.98	100	7.69	40	1.22	6.32	4.89
1994*	71	7.68	39	7.90	95	10.11	38	7.37	103	12.60	37	1.25	6.30	5.88
1995*	74	9.10	33	7.80	96	10.24	36	7.12	109	12.24	41	1.29	6.06	5.53
1996**	69	8.27	32	8.50	101	14.61	21	8.10	107	13.88	20	1.36	6.27	5.97
1997*	73	NA	36	10.00	98	NA	41	7.30	112	NA	12	1.39	7.19	5.25

* States included: winter wheat: CO, ID, IL, KS, MO, MT, NE, OH, OK, OR, SD, TX, and WA; spring wheat: MN, MT, ND, and SD.

** States included: winter wheat: CO, ID, KS, MT, NE, OK, OR, SD, TX, and WA; spring wheat: MN, MT, and ND.

NA = not available.

Sources: Average rate per acre is based on data from the Cropping Practices Surveys and Agricultural Resource Management Survey (ARMS) ARMS. Cost per acre for 1972-85 is computed by multiplying rate per acre (pounds) with the price of seed per pound; cost per acre for 1987-96 is based on Cropping Practices Surveys and ARMS. Seed prices and deflator are from *Agricultural Statistics* (USDA, NASS, various years).

Cotton. Average cotton seeding rates and seed costs per acre vary from year to year. From 1972 to 1985, average seeding rates per acre ranged from 24 to 26 pounds. However, average cost per acre was highest in 1985 ($11.57) due to a higher cotton seed price of $48.20 per 100 pounds. From 1986 to 1996, cotton seeding costs per acre were lower due to lower seeding rates. Although cotton seed price over the same period was much higher than in previous years, lower seeding rates offset the higher cost of producing cotton. From 1986 to 1997, the share of the cotton acres planted with purchased cotton seed ranged from 64 to 81 percent.

Wheat. The average seeding rates per acre for winter wheat vary from year to year (table 7). From 1972 to 1996, the average seeding rates per acre ranged from 66 to 77 pounds. Areas where moisture is plentiful in the growing season, due to either heavy rain or irrigation, support heavier seeding rates. Seeding costs per acre reflect either seed price or seeding rate per acre or both. Average share of winter wheat acres planted with purchased seed also varied from year to year. The type of winter wheat (white, soft red, and hard red), as well as local economic situations, seed prices, and yield of purchased seed, apparently account for much of the variation. For example, in 1993, Washington, Idaho, Oregon, Ohio, and Illinois used purchased seed on more than 60 percent of the winter wheat acreage. In the rest of the producing States, the share of acreage planted from purchased seed ranged from 21 to 51 percent (USDA, 1993). The average seed cost per acre of spring and durum wheat also varied across geographic regions and from year to year due to changes in seeding rate and seed price. Farmers used homegrown spring and durum wheat seed on more than 50 percent of their acreage.

International Seed Markets

The United States is a net exporter of seed, with a seed trade surplus of $384 million in 1996 (table 8). The value of U.S. seed exports grew from $305 million in 1982 to $698 million in 1996. This increase primarily reflects increases in exports of seed for forage crops, vegetables, flowers, and corn. Mexico, Canada, Italy, Japan, France, the Netherlands and, more recently, Argentina, are the largest importers of U.S. seed. Together, these countries accounted for approximately 72 percent of total U.S. seed exports in 1996 (table 9). On a regional basis, North America and Central America (39 percent), Western Europe (29 percent), Middle East and Asia (18 percent), and South America (10 percent) accounted for 96 percent of the total value of U.S. exports in 1996.

Some demand for seed in the United States is met by imports, which compete with U.S.-grown seed in quality, price, and other factors. While both exports and imports of seeds grew substantially over the past two decades, the growth rate of seed imports has exceeded that of seed exports in the United States. The value of total seed imports increased from $87 million in 1985 to $314 million in 1996, a growth rate more than three times that of exports (table 8). This increase is largely reflected in increased seed imports for corn, forage, vegetables, flowers, and other crops. Moreover, the United States also takes part in an extensive amount of multidirectional trade in international seed markets, in which the seeds for many of the same crops are exported and imported.

In 1996, U.S. demand for imported seed was met by a number of different countries. Canada (28 percent), Chile (18 percent), and the Netherlands (9 percent) had the highest shares of total value of U.S. seed imports. On a regional basis, the leading suppliers of seed to the United States were North America and Central America, followed by South America, Asia, and Western Europe (table 10).

In terms of volume, the six leading importing countries of U.S. field corn seed are Italy, Mexico, Canada, France, the Netherlands, and Spain. Shares of these countries' imports of U.S. field corn, which vary from year to year, dropped from 79 percent in 1995 to 63 percent in 1996 (table 11). Over the same period, the volume of U.S. corn seed exports to the six leading importers fell to 53,955 metric tons, a 19-percent decline. Despite this drop, total U.S. corn seed exports to all countries reached 86,183 metric tons in 1996, an increase of 2 percent over 1995 (table 11).

The top three importing countries of U.S. soybean seed by volume are Mexico, Italy, and Japan. In 1996, volume of U.S. soybean seed exports to these three leading trading partners was 63,881 metric tons, an increase of 11 percent from 1995. Together these countries accounted for 86 percent of total U.S. soybean seed exports. From 1995 to 1996, soybean seed exports to Japan increased 41 percent but exports to Italy declined 12 percent and exports to Mexico remained unchanged. Over the same period, however, the volume of total U.S. soybean seed exports to all countries increased 16 percent, from 63,982 metric tons to 73,911 metric tons.

Table 8. Exports and imports of U.S. seed for planting

Item	1982	1983	1984	1985	1986	1987	1988	1989	1990	1991	1992	1993	1994	1995	1996
	Million dollars														
Exports:															
Forage	66	65	70	59	74	75	94	96	104	101	114	116	120	115	139
Vegetables	115	122	135	120	128	138	167	153	176	220	221	218	240	242	231
Flowers	5	6	9	8	9	8	9	11	13	14	19	19	22	27	32
Corn[1]	55	73	46	89	77	63	66	68	138	181	177	168	181	165	144
Grain sorghum	28	32	33	33	29	16	29	55	27	28	34	34	20	12	19
Soybeans	22	12	19	17	19	36	26	54	45	41	30	28	23	25	25
Trees/shrubs	2	2	2	2	2	2	3	4	2	2	3	2	2	2	4
Sugarbeets	4	4	3	2	2	1	2	1	2	3	3	3	4	2	3
Other	8	14	5	28	31	33	26	68	81	82	68	31	41	77	101
Total exports	305	330	322	358	371	372	422	510	588	672	669	619	652	667	698
Deflator, index of prices received by farmers	0.938	0.948	1.000	0.901	0.854	0.867	0.966	1.015	1.017	0.974	0.965	0.991	0.977	0.995	1.097
Exports in 1984 $	325.0	348.2	322.0	397.2	434.6	428.8	436.8	502.3	578.2	690.1	693.6	624.8	667.4	670.1	636.2

Item	1982	1983	1984	1985	1986	1987	1988	1989	1990	1991	1992	1993	1994	1995	1996
	Million dollars														
Imports:															
Forage	17	34	17	18	39	65	52	43	35	31	45	45	39	44	52
Vegetables	31	31	32	34	42	49	58	56	60	79	82	83	85	97	107
Flowers	12	10	18	18	18	21	21	24	23	24	27	31	35	43	56
Corn[2]	12	6	22	14	9	5	10	37	18	15	35	29	44	35	63
Trees/shrubs	1	1	1	1	1	1	2	2	2	2	2	2	1	2	4
Other	3	2	1	2	3	4	4	6	9	14	10	17	23	22	32
Total imports	76	84	90	87	112	146	147	168	147	165	201	208	227	243	314
Imports in 1984 $	81.0	88.6	90.0	96.5	131.2	168.3	152.2	165.5	144.6	169.4	208.4	209.9	232.4	244.1	286.2
Trade balance	229	246	231	271	258	226	275	342	441	507	463	411	425	424	384
Trade balance in 1984 $	244.0	259.6	231	300.6	303.4	260.5	284.6	336.8	433.7	520.6	485.2	414.8	435.1	426.0	350.0

[1] Does not include seed for sweet corn or corn seed provided as food aid.

[2] Certified corn.

Source: Compiled by Mohinder Gill from USDA's Foreign Agricultural Service (FAS) data.

Table 9. Regional and country shares of U.S. seed exports

Region and country	1982	1983	1984	1985	1986	1987	1988	1989	1990	1991	1992	1993	1994	1995	1996
	Percent of total value														
North and Central America	30.0	32.8	31.2	25.0	20.7	24.7	23.0	33.4	26.4	24.8	30.7	33.7	31.7	29.5	39.0
Canada	9.5	8.8	8.9	7.4	6.3	9.4	8.3	6.2	10.2	10.0	11.3	11.8	11.9	13.2	12.2
Mexico	18.0	21.9	19.6	15.1	12.3	13.0	12.6	25.4	14.4	13.2	17.8	20.1	18.0	14.7	25.5
Other	2.5	2.1	2.7	2.5	2.0	2.3	2.1	1.8	1.8	1.6	1.6	1.8	1.8	1.6	1.3
South America	6.9	8.0	8.8	7.0	9.0	7.6	10.9	5.7	4.3	4.1	5.6	7.4	8.1	8.6	9.8
Argentina	0.7	1.0	2.0	1.1	2.5	2.5	3.0	2.0	1.4	1.4	3.0	3.0	4.1	4.4	6.0
Brazil	1.7	2.0	1.0	1.1	1.1	1.2	1.2	0.6	0.5	0.6	0.5	1.0	0.9	1.1	0.8
Chile	0.3	0.6	1.1	0.3	0.4	0.5	0.8	0.6	0.6	0.6	0.6	1.0	1.1	1.0	1.0
Colombia	1.0	1.2	1.4	0.8	1.0	0.9	1.1	0.6	0.6	0.6	0.6	0.7	0.6	0.7	0.6
Venezuela	2.4	1.9	2.2	2.9	3.0	1.4	3.3	0.6	0.6	0.6	0.4	0.7	0.5	0.5	0.4
Other	0.8	1.4	1.0	0.8	1.0	1.1	1.5	1.4	0.5	0.5	0.5	1.1	0.8	0.9	1.0
Western Europe	35.2	29.8	26.4	37.5	36.4	39.7	34.9	29.4	40.7	42.3	36.1	30.3	30.3	34.9	29.3
France	5.5	4.9	4.2	9.6	6.2	4.5	4.5	3.7	7.4	7.2	6.3	6.0	4.9	5.1	4.2
Germany	2.7	2.0	1.7	1.8	1.7	1.6	1.4	1.3	1.7	3.0	1.9	1.8	2.8	1.6	1.9
Greece	2.3	1.7	1.4	1.9	2.3	1.3	1.8	1.0	1.4	2.0	1.3	1.8	2.8	1.9	3.4
Italy	10.2	8.5	8.3	12.5	12.7	19.3	12.4	11.2	16.8	16.2	14.8	10.7	10.0	11.5	10.1
Nether-lands	4.8	4.8	3.4	4.6	5.8	5.4	4.4	4.2	4.9	6.0	4.4	4.7	4.7	6.6	3.7
Spain	1.5	1.0	1.2	1.4	1.6	2.1	4.4	2.9	3.4	2.5	2.4	2.0	2.3	3.7	2.5
United Kingdom	3.4	2.6	2.2	2.7	2.8	2.6	2.9	2.6	1.9	1.7	1.7	1.4	1.4	1.9	2.0

Region and country	1982	1983	1984	1985	1986	1987	1988	1989	1990	1991	1992	1993	1994	1995	1996
	Percent of total value														
Other	4.6	4.2	4.0	3.1	3.4	2.8	3.1	2.5	3.2	3.5	3.5	1.8	1.3	2.6	1.5
Eastern Europe and former Soviet Union	0.9	1.1	0.5	3.3	4.8	0.2	2.0	1.0	2.0	3.7	1.6	7.2	5.4	1.3	0.8
Romania	0.2	0.2	0.2	0.3	0.5	0.0	0.6	0.3	0.8	1.5	0.5	2.0	2.2	0.0	0.0
Ukraine	0.4	0.1	0.0	0.3	0.7	0.0	0.0	0.0	0.0	0.0	0.0	4.7	0.5	0.7	0.1
Other	0.3	0.8	0.5	2.7	3.6	0.2	1.3	0.7	1.2	2.2	1.1	0.6	2.7	0.5	0.7
Middle East and Asia	20.8	21.3	26.5	21.9	23.9	22.6	24.4	26.0	22.3	21.2	21.6	17.1	20.9	19.8	17.6
Hong Kong												0.4	2.6	0.6	0.3
Japan	11.9	11.2	11.5	10.7	9.5	12.3	11.8	8.9	7.8	7.1	8.1	8.2	9.0	8.8	9.8
Saudi Arabia	3.4	3.1	4.5	2.8	3.6	2.0	4.2	10.4	10.0	9.4	7.9	2.3	2.6	1.7	1.5
South Korea	0.5	0.7	1.5	0.8	0.9	1.0	1.0	0.7	0.6	0.7	0.9	1.2	1.7	1.4	1.3
Other	5.0	6.3	9.0	7.6	9.9	7.4	7.5	6.0	3.9	4.0	4.7	4.9	5.0	7.2	4.7
Africa	4.1	4.5	4.7	2.9	3.2	3.0	2.9	2.7	2.7	2.5	2.9	3.0	2.1	3.8	1.3
Oceania	2.1	2.6	1.9	2.4	1.9	2.2	2.0	1.5	1.5	1.4	1.4	1.3	1.7	2.0	1.7
Australia	1.7	2.2	1.5	2.0	1.6	1.8	1.7	1.2	1.2	1.2	1.2	1.0	1.3	1.6	1.4
Other	0.4	0.4	0.4	0.4	0.3	0.4	0.3	0.3	0.3	0.2	0.2	0.3	0.3	0.4	0.3
World[1]	100.0	100.0	100.0	100.0	100.0	100.0	100.0	100.0	100.0	100.0	100.0	100.0	100.0	100.0	100.0

[1] Totals may not add due to rounding.

Source: Compiled by Mohinder Gill from USDA's Foreign Agricultural Service (FAS) data.

Table 10. Regional and country shares of U.S. seed imports

Region and country	1982	1983	1984	1985	1986	1987	1988	1989	1990	1991	1992	1993	1994	1995	1996
	Percent of total value														
North and Central America	30.0	32.8	31.2	38.5	43.4	44.4	35.8	33.2	33.4	31.4	32.6	31.2	33.9	37.1	33.5
Canada	9.5	8.8	8.9	26.7	35.1	37.7	30.4	27.0	26.1	24.9	25.4	22.8	27.2	29.2	27.5
Mexico	18.0	21.9	19.6	4.0	2.9	2.0	2.1	3.2	3.4	3.8	3.2	4.5	3.1	3.8	3.5
Other	2.5	2.1	2.7	7.7	5.4	4.7	3.2	3.2	4.0	2.7	3.9	3.9	3.6	4.1	2.5
South America	6.9	8.0	8.8	9.2	7.0	6.0	8.3	17.7	12.3	11.7	21.2	17.2	18.9	17.4	24.0
Argentina	0.7	1.0	2.0	0.0	0.0	1.5	0.7	4.0	1.0	0.7	3.6	3.3	2.8	2.0	4.9
Chile	0.3	0.6	1.1	8.2	6.2	4.0	6.8	13.3	11.2	10.7	17.2	13.5	15.7	14.8	18.1
Other	5.9	7.0	5.6	1.0	0.8	0.5	0.8	0.4	0.1	0.3	0.4	0.5	0.4	0.6	1.0
Western Europe	35.2	29.8	26.4	21.0	18.7	19.2	16.8	16.3	15.7	18.0	16.2	21.7	18.7	18.1	15.9
Denmark	1.1	0.9	0.7	1.6	1.2	2.1	1.9	1.5	1.4	1.2	1.7	2.6	1.3	1.5	1.8
France	5.5	4.9	4.2	1.4	1.1	1.7	1.0	1.3	1.4	1.8	1.7	2.2	2.4	2.3	2.0
Germany	2.7	2.0	1.7	1.4	2.2	2.5	2.2	1.8	1.6	2.2	2.0	2.5	2.4	1.9	2.0
Italy	10.2	8.5	8.3	1.7	1.1	1.2	1.6	1.0	1.0	1.2	1.2	0.9	0.9	1.1	0.9
Netherlands	4.8	4.8	3.4	11.7	10.5	10.2	9.2	8.9	9.3	10.7	8.7	12.3	10.6	10.2	7.8
Other	10.7	8.6	8.0	3.2	2.5	1.5	0.9	1.9	0.9	1.0	0.9	1.2	1.2	1.1	1.4
Eastern Europe and former Soviet Union	0.9	1.1	0.5	1.8	0.4	0.5	1.3	3.3	0.7	0.1	0.5	1.7	1.9	0.5	0.4

Region and country	1982	1983	1984	1985	1986	1987	1988	1989	1990	1991	1992	1993	1994	1995	1996
	Percent of total value														
Asia	20.8	21.3	26.5	20.1	22.2	19.4	24.5	24.9	30.5	31.3	22.4	21.1	20.0	19.9	21.1
China				0.0	0.0	0.0	2.4	3.9	6.8	9.0	5.5	4.8	6.0	7.0	7.3
India	0.0	0.0	0.0	3.3	6.5	2.9	7.5	3.5	3.1	5.2	2.7	1.8	1.9	2.4	2.4
Japan	11.9	11.2	11.5	6.1	6.1	6.0	6.4	6.9	7.6	7.8	6.2	9.0	7.7	6.3	5.3
Taiwan	0.0	0.0	0.0	7.6	6.0	6.7	4.5	6.2	4.8	3.5	3.6	1.8	0.8	1.1	1.0
Thailand	0.0	0.0	0.0	0.0	0.0	0.7	1.6	2.8	5.4	5.3	4.0	2.8	2.0	2.5	2.6
Other	8.9	10.1	15.0	3.0	3.6	3.1	2.0	1.5	2.9	0.5	0.4	0.8	1.5	0.6	2.5
Africa	4.1	4.5	4.7	6.4	4.0	3.9	3.8	0.7	0.9	1.3	1.1	1.4	1.0	1.6	1.0
Oceania	2.1	2.6	1.9	3.0	4.3	6.5	7.4	4.0	6.4	5.5	4.5	4.6	3.1	3.3	3.5
Australia	1.7	2.2	1.5	2.2	1.8	2.1	1.8	1.6	2.2	1.9	2.0	1.4	1.9	1.6	1.6
New Zealand	0.4	0.4	0.4	0.8	2.6	4.5	5.6	2.4	4.2	3.5	2.5	3.2	1.2	1.7	1.9
Other	0.0	0.0	0.0	0.0	0.0	0.0	0.0	0.0	0.0	0.0	0.0	0.0	0.0	0.0	0.0
World[1]	100.0	100.0	100.0	100.0	100.0	100.0	100.0	100.0	100.0	100.0	100.0	100.0	100.0	100.0	100.0

[1] Totals may not add due to rounding.

Source: Compiled by Mohinder Gill from USDA's ForeignA gricultural Service (FAS) data.

Table 11. U.S. exports of corn and soybean seed by country of destination and volume

Commodity/country	1986	1987	1988	1989	1990	1991	1992	1993	1994	1995	1996
	Metric tons										
Corn seed:											
Canada	1,621	2,505	2,582	1,548	4,076	7,561	8,192	8,678	15,421	16,890	13,217
France	2,121	2,542	2,439	2,873	9,666	10,953	13,859	8,177	6,188	7,732	5,030
Italy	7,939	12,229	8,741	12,168	20,889	21,773	23,955	14,178	11,782	15,042	10,383
Mexico	3,703	3,143	3,151	10,205	10,329	7,963	12,472	15,750	16,453	15,175	15,088
Netherlands	5,127	695	1,060	351	2,437	10,354	2,791	4,834	7,977	7,062	6,691
Spain	1,245	2,049	4,134	1,836	4,132	2,076	2,853	2,492	4,004	4,796	3,546
Subtotal	21,756	22,121	22,915	28,981	51,529	66,952	65,054	56,039	62,170	66,697	53,955
All others	22,906	10,291	10,632	7,876	18,837	26,770	13,455	32,690	43,126	17,724	32,228
Total, all countries	44,662	32,412	33,547	36,857	70,366	93,722	78,509	88,729	105,296	84,421	86,183
Soybean seed:											
Italy	NA	44,348	27,833	20,185	55,937	65,571	34,500	27,764	15,711	14,672	12,970
Japan	NA	4,151	5,277	1,608	2,325	6,947	7,341	7,439	9,654	19,741	27,751
Mexico	NA	12,630	8,922	100,380	36,731	4,827	32,674	19,472	8,891	23,116	23,160
Subtotal	NA	61,129	42,032	122,173	94,993	77,345	74,515	54,675	34,256	57,572	63,881
All others	NA	14,035	11,698	6,409	11,998	13,659	9,070	14,216	13,700	6,410	10,030
Total, all countries	NA	75,164	53,730	128,582	106,991	91,004	83,585	68,891	47,964	63,982	73,911

NA = not available.

Source: Compiled by Mohinder Gill from USDA's Foreign Agricultural Service (FAS) data.

REGULATIONS HAVE AFFECTED THE SEED INDUSTRY

Agricultural innovations, such as improved plant varieties, are a product of research and development. Seeds embody the scientific knowledge needed to produce a new plant variety with desirable attributes, such as higher yield, greater disease resistance, or improved quality. To fully understand the nature of the seed industry, it is necessary to consider the regulations that affect the costs and benefits faced by public and private sector innovators, agricultural producers, and other agents in the seed market.

Appropriability and Agricultural R&D

Some agricultural innovations are imperfectly appropriable, meaning that the innovation, or the knowledge embodied in the innovation, can be transmitted to, imitated by, or reproduced by prospective competitors with minimal difficulty or at a low cost, and with little or no obligation to compensate the innovators (Cohen and Levin, 1989, pp. 1090-1991). Plant breeders, in particular, face both the risk of imitation by competing seed firms and the risk of seed reproduction by farmers themselves. For example, once marketed, plant breeding innovations embodied in the seeds of improved self-pollinated varieties, such as wheat, can be easily adapted by competing seed firms into their own product lines without compensation to the innovators if property right protections are not available (Beach and Fernandez-Cornejo, 1994, p. 5). Once the seeds are sown, they can also be reproduced and used by farmers as seed for planting in subsequent years, again without compensation.

If innovators are unable to assert property rights over their innovations or the knowledge used in creating innovative products, they may be unable to realize the full rewards of their efforts. This effect may reduce the private incentives for further innovations. If the innovation provides social benefits, as is frequently the case with agricultural sector innovations, then limited private incentives may result in research underinvestment. The establishment of patent laws and other forms of enforceable legal protection, which provide innovators limited market power, thereby generating private incentives for research, offer a potential solution to this appropriability problem and its social consequences. Public investments in socially desirable research and development, particularly in areas in which private incentives are inadequate, offer another possible solution.

Timeline of Regulations Related to Intellectual Property Rights to Plant Varieties

1790. The first intellectual property rights (IPR) legislation passed by Congress was the ***Patent Act of 1790***, which protects the intellectual property rights of inventors, discoverers, and innovators and establishes a framework through which these individuals may obtain financial rewards through a functioning market system. The Patent Act and its subsequent amendments do not, however, extend IPRs to new plant varieties; rather, they classify biological innovations, such as new plant varieties, as "products of nature" and exclude them from protection.

1883. One of the oldest international IPR agreements is the ***Paris Convention for the Protection of Industrial Property*** of 1883, which seeks to harmonize patent regimes among its 100 signatory countries. However, the convention provides its members only limited property rights protection for innovation of plant varieties and biological processes for plant production.

1930. The first IPR legislation enacted to specifically address issues of plant breeding was the ***Plant Patent Act of 1930 (PPA)***. Administered by the U.S. Patent and Trademark Office (PTO), the PPA provides patent protection over asexually or vegetatively reproduced plant varieties. The PPA also includes patent protection for spores, mutants, hybrids, newly found seedlings, or plants found in an uncultivated state, and extends property rights for a period of 17 years.

1952. The ***Patent Act of 1952 (PA)*** extends patent rights to agricultural innovations under a much more general category that includes "any new and useful process, machine, manufacture, or composition of matter, or any new and useful improvements thereof." Patent protection under the PA covers agricultural machinery, equipment, chemicals, production processes, and similar inventions, and is termed "utility patent protection." More importantly, the PA's broad definition of what may be entitled to patent protection leaves an important opening for covering innovations in biotechnology and genetic engineering.

1961. The International Convention for the Protection of New Varieties of Plants was adopted in Paris, France, with the objective of providing intellectual property rights to the breeders of new varieties of plants.

The Convention established the ***International Union for the Protection of New Varieties of Plants*** (UPOV—*Union Internationale pour la Protection des Obtentions Végétales*), an intergovernmental organization headquartered in Geneva, Switzerland. The Convention was revised in 1972, 1978, and 1991. The 1991 revision attempts to expand protection to address new issues in agricultural biotechnology.

1960s. The American Seed Trade Association formed the Breeders' Rights Study Committee to examine issues related to plant breeders' property rights. This effort helped enact the Plant Variety Protection Act in December 1970.

1970. *The Plant Variety Protection Act (PVPA)* grants breeders a Certificate of Protection that gives them exclusive rights to market a new plant variety for 18 years from the date of issuance. These exclusive rights are subject to a research exemption and a farmer's exemption.

1971. The Consultative Group on International Agricultural Research (**CGIAR**) is established as a key institution in the free international exchange of plant genetic materials. CGIAR, a global network of agricultural research centers, receives funding from multilateral agencies, governments of both industrialized and developing countries, and private foundations. Included within CGIAR's charter is the coordination of efforts to preserve plant genetic material and distribute these resources to research institutions in member countries.

1980. Breeders' rights were strengthened by the U.S. Supreme Court's 1980 decision in ***Diamond*** v. ***Chakrabarty***, which extends patent rights to genetically engineered microorganisms, an important tool and product of biotechnology.

1983. FAO member countries passed Resolution 8/83, the International Undertaking on Plant Genetic Resources (the Undertaking), to ensure free access to genetic material whether existing in the public domain or developed commercially.

1985-88. A series of rulings by the PTO's Board of Appeals and Interferences widened the scope of patent protection for genetically engineered organisms by including plants and nonhuman animals. These rulings extend IPR to a wide range of new biotechnology products in the form of utility patents awarded under the PA. These products include seeds, plants, plant parts, genes, traits, and biotechnology processes.

1994. The ***1994 amendment to the PVPA***, which went into effect in April 1995, brought the PVPA into conformity with international standards established by the International Union for the Protection of New Varieties of Plants and allowed the United States to ratify the 1991 International Convention for the Protection of New Varieties. Protection provided by Certificates of Protection extended from 18 to 20 years for most crops.

2000. A case involving Pioneer Hi-Bred brought before the U.S. Federal Court of Appeals reinforced plant breeders' intellectual property protection through protection certificates issued under the PVPA or through utility patents awarded under the PA. This ruling extended the options available to plant breeders seeking to assert property rights over their innovations.

2001. FAO members approved an International Treaty on Plant Genetic Resource for Food and Agriculture growing out of the International Undertaking in November 2001, although the agreement is subject to ratification by member states.

IPR Protection in the Seed Industry

Providing private incentives to innovators and inventors is a clear and longstanding priority in U.S. agriculture and industry. The U.S. Constitution charges the Congress with the responsibility of establishing laws that award innovators exclusive proprietary rights over their inventions and ideas for limited periods of time. The first intellectual property rights (IPR) legislation passed by Congress was the Patent Act of 1790, which protects the intellectual property rights of inventors, discoverers, and innovators, and establishes a framework through which they may obtain financial rewards through the functioning of the market system (see box on timeline of regulations). The Patent Act and its subsequent amendments, however, do not extend IPRs to new plant varieties; rather, they classify biological innovations, such as new plant varieties, as "products of nature" and exclude them from protection (Fuglie et al., 1996, p. 35).

In 1790, the lack of protection over plant varieties was of limited relevance to most farmers, breeders, and other agricultural sector participants because farmers of that era relied on nonhybrid varieties of seed for planting new crops. These plant varieties were seeded by the natural processes of pollination. Seeds from self-pollinating crops, such as wheat or cotton, could be saved from one crop harvest and planted for the next without the seed losing yield potential or vigor. Seeds from cross-pollinating crops, such as corn, could also be saved from one harvest and planted for the next. However, before the advent of hybrid varieties, farmers had to select more carefully among these seeds to maintain the desirable characteristics they wanted.[1]

The use of saved seeds to plant subsequent crops severely limited the extent to which innovators might realize the benefits of plant breeding efforts. In practical terms, it was nearly impossible for an innovator to maintain appropriability over nonhybrid seeds, thus limiting the role of an IPR regime where nonhybrid seeds play a central role in agriculture.

By the 1920s, the development of hybrid corn seed offered farmers an alternative to open-pollinated corn. Hybrids also proved beneficial to plant breeders: As long as the lineage of a hybrid remains known only to the breeder, the hybrid cannot be easily reproduced, thus providing the plant breeder with control and appropriability over the innovation. Moreover, seed saved and planted from the harvest of a hybrid crop tends to diminish in yield and vigor in subsequent harvests, thus ensuring breeders a continuous market for their seed so long as other higher performing hybrid seeds do not enter the market. The unique nature of hybrids led to extensive commercialization of the corn seed industry in the 1930s, even in the absence of a regulatory framework to protect new plant varieties.

The first IPR legislation passed to specifically address issues of plant breeding was the Plant Patent Act of 1930 (PPA). Administered by the U.S. Patent and Trademark Office (PTO), the PPA provides patent protection over asexually or vegetatively reproduced plant varieties. PPA specifically covers plants derived from parts of the parent other than its seeds or tubers, and thus covers plants that contain the exact genetic makeup as the parent plant. The protection includes spores, mutants, hybrids, newly found seedlings, or plants found in an uncultivated state and extends property rights for a period of 17 years (USITC, 1995, p. 16). Patent owners have the right to exclude others from reproducing their plants asexually or vegetatively and may enforce ownership through civil action brought against parties alleged to be infringing upon their patents. The PPA's explicit exclusion of plants that are sexually reproduced or propagated by tubers reflects the perception at the time that such varieties were not adequately identifiable, uniform, or stable enough to constitute varieties requiring patent protection.

The Patent Act of 1952 (PA) extended patent rights to agricultural innovations under a much more general category, which includes "any new and useful process, machine, manufacture, or composition of matter, or any new and useful improvements thereof" (35 U.S.C. § 101, in USITC, 1995, p.

16). The PA also includes patent protection for agricultural machinery, equipment, chemicals, production processes, and similar inventions and is termed "utility patent protection." The protection and exclusionary rights offered under the 1952 PA are, under many circumstances, significantly greater than similar protections and rights offered by the 1930 PPA and the 1970 Plant Variety Protection Act. More importantly, the PA's broad definition of what may be entitled to patent protection left an opening for covering innovations in biotechnology and genetic engineering.

Still, neither the 1930 PPA nor the 1952 PA contained language that extended IPR to seed- and tuber-propagated varieties. In the early 1960s, the American Seed Trade Association (ASTA) formed the Breeders' Rights Study Committee to examine the issue, an effort that contributed to the enactment of the Plant Variety Protection Act (PVPA) in December 1970 (USITC, 1995, p. 16).

The 1970 PVPA grants breeders a Certificate of Protection that gives them exclusive rights to market a new variety for 18 years from the date of issuance. These exclusive rights are subject to two exemptions: (i) a research exemption, which allows the use of the seed to develop new varieties; and (ii) a farmer's exemption, which allows a farmer whose primary occupation is growing crops for sale to save seed from a protected variety to plant on the farmer's land, and to sell from that seed to another farmer whose primary occupation also is to grow crops (Strachan, 1992). Saved seed that is sold under exemption (ii) must meet the applicable State seed laws and must be labeled to show the variety name as protected under the PVPA. Further, the PVPA does not extend protection to fungi, bacteria, and first-generation hybrids. Like the 1930 PPA, the 1970 PVPA is enforceable through the actions of a protected variety's owner. To enforce protection, an owner of a protected variety may bring civil action against parties allegedly infringing on his/her property rights, and would typically seek an injunction to prevent others from further violations (USDA, AMS, 2000a).

The 1994 amendment to the PVPA, which went into effect in April 1995, brought the PVPA into conformity with international standards established by the International Union for the Protection of New Varieties of Plants (UPOV)[2] and allowed the United States to ratify the 1991 International Convention for the Protection of New Varieties (Fuglie et al., 1996, p. 35). The amendment extends the length of protection provided by a Certificate of Protection from 18 to 20 years from the date of issuance for most crops (25 years for trees, shrubs, and vines) (USDA, AMS, 2000a). The amendment also prohibits farmers from selling saved seed of protected varieties without the permission of the variety owner (Fuglie et al., 1996, p. 35). In addition, the amendment extends protection to tuber-reproduced plants (such as potatoes). The amendment's introduction of an "essentially derived" plant variety category, which entitles such varieties to protection, is specifically designed to address technological advances made in biochemistry and genetic engineering that enable breeders to develop varieties that may differ on the basis of a single gene or micro-molecule within the DNA structure. The category extends the definition of distinctness to include varieties of GE plants for which the uniqueness exists at miniscule levels, thereby providing property rights to plant breeders over even the smallest of genetic manipulations of their varieties (USITC, 1995, p. 16; USC, 1970, § 2401, 2541).

The PVPA affords IPR to plant varieties that are demonstrably "new, distinct from other varieties, and genetically uniform and stable through successive generations" and includes protection for both nonhybrid and hybrid seeds (USDA, AMS, 2000a).[3] According to the PVPA, distinctness, a key determinant of a variety's potential for protection, may be based on "one or more identifiable morphological, physiological, or other characteristics (including any characteristics evidenced by processing or product characteristics, such as milling and baking characteristics in the case of wheat) with respect to which a difference in genealogy may contribute evidence" (USDA, AMS, 2001a).

The PVPA is administered by USDA's Plant Variety Protection Office (PVPO). The PVPO is responsible for scrutinizing applications for Certificates of Protection, including information on the variety's lineage, genealogy, and breeding methodology, as well as seed or cell-culture samples and other proof of the variety's distinctness, uniformity, and stability. Plant breeders applying for protection of new wheat varieties must also submit information on the milling and baking characteristics of the variety (USDA, AMS, 2000a). Applications may be submitted by both domestic and foreign breeders seeking protection for their variety in the U.S. marketplace.

Breeders' rights were strengthened by the U.S. Supreme Court's 1980 decision in *Diamond* v. *Chakrabarty*, which extends patent rights to genetically engineered microorganisms, an important tool and product of biotechnology. In the case brought before the Supreme Court, the underlying question was whether a genetically engineered bacterium designed to digest and break down crude oil was a "product of nature" that was not covered by the Patent Act or whether it was a new invention for which a patent could be awarded. Among the arguments brought before the Court was the fact that patents had been previously awarded for compositions containing living organisms, such as microbial spores, vaccines, yeast compositions, and certain dairy products. Ultimately, the Supreme Court determined that GE microorganisms were, in fact, patentable (Schor, 1994, pp. 60-61). A series of rulings by the PTO's Board of Appeals and Interferences widened the scope of patent protection for genetically engineered organisms by including plants and nonhuman animals. These rulings extend IPR to a wide range of new biotechnology products in the form of utility patents awarded under the PA. Products protected under the rulings include seeds, plants, plant parts, genes, traits, and biotechnology processes (Fuglie et al., 1996, p. 35; USITC, 1995, p. 16).

Breeder's rights were extended further with the Supreme Court's 1995 decision in *Asgrow* v. *Winterboer*, which precluded farmers from selling protected seed without a license from the variety's owner for varieties developed before April 1995 and not covered by the PVPA's 1994 amendment. This decision, along with the PVPA amendment, addressed the issue of appropriability in terms of the threat posed to plant breeders not by competing firms but by farmers who save and reproduce nonhybrid seed from their own crops for resale purposes (Fuglie et al., 1996, p. 35).[4]

A more recent case involving Pioneer Hi-Bred brought before the U.S. Federal Court of Appeals in 2000 reinforced plant breeders' intellectual property protection through protection certificates issued under the PVPA or through utility patents awarded under the PA (*AgBiotech Reporter*, 2000). This ruling extended the options available to plant breeders seeking to assert property rights over their innovations.

Both legislative and judicial action have contributed to an IPR regime in the United States that provides an extensive set of incentives to developing new plant varieties: plant patents for asexually or vegetatively propagated varieties under the PPA; certificates of protection for sexually or tuber-propagated varieties under the PVPA; and utility patents under the PA. Although the number of plant patents issued under the PPA exceeds all other types of property protection awarded to plant innovators, the number of protection certificates and utility patents has increased significantly in recent years (Fuglie et al., 1996, pp. 36-37).

IPR in the International Context

Although the U.S. IPR regime provides a comprehensive framework to protect plant breeders' rights and create incentives for plant breeding R&D,

the relevance of the U.S. regime is better understood within an international context because the expansion of U.S. IPR has implications for genetic resource conservation worldwide. Historically, the United States and several other countries have facilitated the free exchange of plant genetic resources for research purposes as a means of increasing worldwide agricultural production and food security. However, the role of intellectual property rights in this context remains unclear and has been the subject of much international debate (USDA, ERS, 2000, p. 14).

A key institution in the free global exchange of plant genetic materials is the Consultative Group on International Agricultural Research (CGIAR), an international network of agricultural research centers established in 1971. CGIAR receives funding from multilateral agencies, governments of both industrialized and developing countries, and private foundations. Included within CGIAR's charter is the coordination of efforts to preserve plant genetic material and distribute these resources to research institutions in member countries. Historically, the free exchange of plant genetic resources has been important to the United States, not only because of its membership in CGIAR but also because of its need for access to genetic materials beyond U.S. borders. The relative lack of genetic diversity among indigenous plants makes the United States a "germplasm-deficient" country, and free exchange ensures the United States continued access to genetic resources from other countries to support its extensive work in agricultural R&D (Day-Rubenstein and Heisey, 2001, p. 22). In fact, as a result of collection and breeding activities, the United States is currently a net supplier of plant germplasm to the rest of the world (Day-Rubenstein and Heisey, 2001, p. 18).[5]

The United States, however, is also committed to supporting plant breeders and private sector investment in plant breeding R&D, a commitment that is shared with many other countries. As a result, the United States is party to a number of international agreements and conventions designed to protect the rights of plant breeders. One such agreement, the Paris Convention for the Protection of Industrial Property (1883), seeks to harmonize patent regimes among its 100 signatory countries. However, the Paris convention provides only limited property rights protection for plant varieties and biological processes for plant production (Van Wijk, 1993, p. 17). The UPOV provides for a more explicit IPR regime to its 52 member states by extending protection to distinct, uniform, and stable plant varieties for a minimum of 15 years.[6] The 1991 Act of the UPOV convention attempts to expand protection to address new issues in agricultural biotechnology. For instance, the 1991 Act eliminates an exemption for essentially derived varieties, under which breeders who created new varieties by incorporating single genes into an existing protected variety did not require permission from the variety owner (Van Wijk, 1993, pp. 6-7). Out of 52 member states, only 23, including the United States, have become a party to the 1991 Act (UPOV, 2003).

The difficulty of balancing free access to plant genetic materials with protecting breeders' rights was apparent in 1983, when member countries of the United Nations Food and Agriculture Organization (FAO) passed Resolution 8/83, the International Undertaking on Plant Genetic Resources (the Undertaking), seeking to ensure free access to genetic material, whether existing in the public domain or developed commercially. As a result of objections from the United States and other FAO member countries, compliance with the resolution was deemed nonbinding on members. Disagreements arose during subsequent rounds of discussion on key issues related to plant breeding, such as compensating farmers, particularly in developing countries, for their contribution to past plant genetic improvements; protecting the rights of plant breeders over their inventions, ideas, and products; ensuring free and equitable access to genetic materials; and establishing programs to preserve diverse genetic resources for future use. Members approved an International Treaty on Plant Genetic Resource for Food and Agriculture growing out of the Undertaking in November 2001, although it is subject to ratification by member states (FAO, 2001).

The UN Convention on Biological Diversity (1992) also addressed issues relevant to plant breeding, such as the equitable use and preservation of plant genetic resources, although the convention's provisions relating to IPRs were found wanting by the United States and other members (Day-Rubenstein and Heisey, 2001, pp. 20-21; Van Wijk, 1993, pp. 26-27). The convention was signed by the United States in 1993 but has not been ratified by the U.S. Senate (Day-Rubenstein and Heisey, 2001, p. 22).

While FAO, the UN, and CGIAR efforts focus on preserving genetic diversity and, to the extent possible, making plant genetic material available worldwide, a new international regulatory regime is poised to establish a much stricter international IPR regime. The 1986 Uruguay Round of the General Agreement on Tariffs and Trade (GATT) established the framework for an initiative on trade-related aspects of intellectual property rights (TRIP). Under TRIP, member countries of the World Trade Organization (WTO) are required to update their IPR legislation to meet new international standards. These new standards include the protection of seed and plant varieties with patents or similar property rights (Day-Rubenstein and Heisey, 2001, p. 22). Moreover, the WTO's authority to sanction members for noncompliance with TRIP would enable the organization to more effectively enforce the initiative, resulting in a stronger international IPR regime that will reflect and support the present IPR regime in the United States.

Regulation to Ensure Seed Quality

Along with the comprehensive framework designed to protect plant breeders' rights, the United States also offers protections to farmers who purchase seed, directly or indirectly, from plant breeders. Because the quality of most seed cannot be determined by visual inspection, the risks associated with seed choice are high. U.S. farmers are protected by a comprehensive system to ensure seed quality.

Varietal registration, a key protection for farmers who purchase seed, provides a system for establishing a variety's genetic identity and its performance characteristics, such as yield or disease resistance. In the United States, plant breeders register varieties with Federal agencies responsible for awarding plant patents or protection certificates. These agencies can provide farmers with information on the characteristics of different varieties that might otherwise be overly complicated or difficult to obtain in the marketplace. U.S. plant breeders are not required to provide information on a variety's performance characteristics, which is typically ascertained through field tests (Tripp, 1998, p. 160). Field testing is mandatory, however, in cases involving the introduction of GE organisms.

Seed certification and quality testing also offer protection to farmers. Seed certification establishes the genetic purity of a seed, while quality testing ascertains such information as germination rates, moisture content, or seed size. Individual States oversee the process of seed certification through State agencies, such as agricultural extension services; State agricultural departments; or independent bodies, such as crop improvement associations. The certification system is not a rigorous process of mandatory testing; rather, seed companies are required to adhere to truthful labeling provisions that permit companies to sell seed as long as seed quality information is completely disclosed on the packaging. This labeling provision is considered highly effective (Tripp, 1998, p. 164).

Environmental and Consumer Protection

Much like the laws that protect plant breeders' property rights and the interests of farmers, a regulatory framework provides protection for the environment and for consumers of agricultural commodities. These laws are particularly relevant in light of the expanding role of biotechnology in U.S. agriculture.

The Animal and Plant Health Inspection Service (APHIS) of the U.S. Department of Agriculture plays a central role in regulating the release of agricultural biotechnology products into the environment. Such products, which include genetically engineered plants, microorganisms, and invertebrates, are considered "regulated articles." Private firms and public institutions wishing to move or release these organisms must receive authorization from APHIS through either a notification or permit procedure. APHIS requires that the notifications and permit applications contain specific details about the organism's genetic makeup and lineage, as well as the testing and safety measures designed to prevent the organisms from being disseminated from the test site or persisting beyond the duration of the test. In the case of permits, APHIS often imposes additional conditions to ensure confinement. APHIS determines whether to authorize the test, based on whether the release will pose a risk to agriculture or the environment. APHIS and State authorities maintain a continuing right to inspect test sites at any time (USDA, 2000b). After years of field tests, an applicant may petition APHIS for a determination of nonregulated status in order to facilitate commercialization of the product. If, after extensive review, APHIS determines that the unconfined release does not pose a significant risk to agriculture or the environment, then the organism is "de-regulated." At this point, the organism is no longer considered a regulated article and can be moved and planted without APHIS authorization.

APHIS is also responsible for plant quarantines, a function that is crucial to protecting the environment from the spread of disease or pests. APHIS enforces regulations that govern the import and export of plants and seeds and are designed to ensure that sanitary and phytosanitary threats do not affect U.S. agriculture or the agriculture sectors of U.S. trading partners. The agency is also responsible for imposing quarantines on areas within the United States where disease or pests pose a threat. APHIS authority to impose quarantines is particularly important with the increase in adoption of bioengineered crops and concerns over genetic exchanges among these crops, weeds, and other crops (Tripp, 1998, p. 171).

If a plant is engineered to produce a substance that "prevents, destroys, repels, or mitigates a pest," then such substance is considered to be a pesticide and is subject to regulation by the U.S. Environmental Protection Agency (EPA) (Federal Register, November 23, 1994). The U.S. Food and Drug Administration (FDA) maintains regulatory control over all food applications of crops, including those crops that are developed through the use of biotechnology. Shoemaker et al. (2001, pp. 31-32) describe the EPA and FDA regulation of agricultural biotechnology products.

SEED INDUSTRY STRUCTURE IS CHARACTERIZED BY GROWTH AND CONSOLIDATION

From the first ventures into the commercial production of hybrid corn seed in the 1930s, to the recent mergers and acquisitions, the seed industry has experienced extensive structural change and transition.

Early Industry Structure: 1920-1970

Until the late 19th century, most U.S. farmers depended on seed saved from their own crops cultivated in the previous year and did not purchase significant quantities of seed from commercial sources. It was not uncommon for farmers to share surplus seed with friends and neighbors. The advent and expansion of seed certification programs between 1915 and 1930 brought about large increases in the number of farmers who purchased seed from commercial traders instead of producing it themselves or obtaining it locally from neighbors. Seed certification programs provided quality assurances to farmers, leading to a rise in the role of commercial seed markets.

Most commercial seed suppliers at that time were small, family-owned private businesses lacking the financial resources necessary to pursue their own R&D activities. The primary role of seed businesses at the time was to multiply and sell seeds of varieties developed in the public domain, as R&D of improved plant varieties was carried out almost exclusively by land-grant colleges and universities, State agricultural experimental stations, and other public agencies (Duvick, 1998; McMullen, 1987).

At the end of the 19th century, the seed used in corn (the dominant field crop in U.S. agriculture) was almost entirely based on open-pollinated varieties (OPV) that farmers saved from prior crops and subsequently planted (Schor, 1994, p. 35). In the early part of the 20th century, public researchers developed high-yielding hybrid corn varieties that consistently outperformed OPVs. Capitalizing on these breakthroughs and the growing demand for hybrid seeds (and given the implicit form of proprietary rights enjoyed by hybrid corn breeders over their innovations), the private sector's role in the commercial market for hybrid corn seed increased significantly beginning in the 1930s (Duvick, 1998, pp. 198-200).

The development and diffusion of hybrid corn varieties, with their inherent capacity to protect returns to private investment, transformed the U.S. seed industry. Beginning in 1930, approximately 150 companies formed to produce hybrid corn seed and some 40 existing seed companies expanded their businesses to include production of hybrid corn seed. While most firms were established to produce and sell seed, some also instituted inhouse research and breeding programs to improve existing hybrids. As long as the lineage of a company's hybrid remained unknown to competitors or farmers, the company continued to hold a unique and marketable product until an even better hybrid was developed. By 1944, U.S. sales in the seed corn market had expanded to over $70 million, establishing corn seed as the core business of the U.S. seed industry (Duvick, 1998, p. 199).

The early growth of the seed industry shifted corn production to hybrids swiftly and extensively; by 1965, over 95 percent of American corn acreage was planted with hybrid seed. Industry expansion also generated profits sufficient to support reinvestment in plant breeding R&D, leading to continual increases in corn seed productivity and crop yields (McMullen, 1987, p. 89). By constantly improving their products through new research, private seed firms were able to maintain the corn seed market's longrun viability. The seed industry reshaped itself primarily around large firms highly vested in the corn seed industry. The smaller firms in the industry tended to be family-run, regionally oriented firms active only in producing, distributing, and marketing varieties developed by the public sector or larger private companies (Kimle and Hayenga, 1993, pp. 19-20). The ability of farmers to save nonhybrid seeds limited the expansion of the seed industry into other agricultural seed markets, establishing corn as the historical force behind the growth of the seed industry.

Modern Industry Structure: 1970-Present

With the exception of hybrid seed firms, few companies had proprietary rights over the plant varieties they sold as seed until the early 1970s.[7] Most private seed firms focused primarily on cleaning, handling, storing, packaging, and selling seed developed in the public domain. The 1970 PVPA, subsequent amendments and rulings, and other actions strengthened property rights by providing proprietary rights over sexually- and tuber-propagated new plant varieties, creating an incentive for private firms to enter the seed market.

Over the past three decades, the U.S. seed industry has been marked by transition. As recently as 1970, most seed firms were independent. During the 1970s, most small seed firms vanished, as mergers and acquisitions created a new seed industry structure dominated by large companies with primary investments in related sectors. For example, more than 50 seed companies were acquired by pharmaceutical, petrochemical, and food firms following the passage of the 1970 PVPA (Lesser, 1998). The acquiring companies were drawn to the potential profits available through the purchase of strong, well-developed seed companies. Those large corporations, many of them multinational conglomerates, possessed the resources needed to achieve scale economies in research and development. Many chemical firms entered the U.S. seed market because the agricultural chemicals market had reached maturity and profits in that sector were declining (Kimle and Hayenga, 1993, pp. 20-21).[8] Pursuing new, high-growth opportunities, large multinational corporations specializing in chemicals and pharmaceuticals, such as Ciba-Geigy, Sandoz, Royal Dutch/Shell, Upjohn, and Celanese, entered the seed industry in the mid-1970s (Kimle and Hayenga, 1993, pp. 19-20). As a result, private sector acquisitions expanded rapidly, and, by the early 1980s, several international firms were among the top seed sellers worldwide (table 12).

In the early 1980s, developments in biotechnology created an additional incentive for firms to increase their R&D capacity and expand further into seed production. As the first products of crop biotechnology began large-scale extensive testing in the 1980s, the seed industry's structure underwent additional transformation. The industry again reorganized through extensive mergers, acquisitions, and joint ventures as companies sought to achieve economies of scale to offset the high costs of biotechnology R&D. Strong demand complementarities provide the rationale for joint ventures between chemical and seed businesses (Just and Hueth, 1993). An example is the case of the herbicide glyphosate and soybeans tolerant to glyphosate.

Table 12. Global seed sales of top international seed companies

Company	1983	1989	1983	1989
	Million current dollars		*Million 1989 dollars*[1]	
Royal Dutch/Shell	650		784	
Pioneer Hi-Bred	557	840	672	840
Sandoz	319	471	385	471
Cardo	285		344	
Asgrow		270		270
DeKalb/Pfizer	187	205	226	205
ICI		250		250
SICA France Mais		170		170
Takii		170		170
Clause		159		159
Claeys-Luck	155		187	
Sakata		152		152
Upjohn	139		168	
Limagrain	130	268	157	268
Ciba-Geigy	107	148	129	148
Suiker Unie	100		121	
K.W.S.	80		97	
Cebeco	65		78	
Svalof	55		66	
Cargill	50	241	60	241

[1] Calculated using the U.S.GDP deflator.

Source: McMullen (1983), p.94; Kimleand Hayenga (1989), p.21.

Despite these incentives, many large chemical and industrial manufacturing companies that invested heavily in the seed industry during the early 1980s are no longer in the seed business. Royal Dutch/Shell, a market leader in 1983 with seed sales topping $650 million, sold its seed unit and had exited the seed industry completely by 1989. Other large players in the market, such as Occidental Petroleum, Upjohn, Lubrizol, and Celanese, similarly shed their seed subsidiaries. Of the 14 companies that led industry sales in 1983, only 7 occupied top global sales positions by 1989. Pioneer Hi-Bred maintained the leadership position in the market in 1989, followed by Sandoz, Asgrow, and Limagrain.

Mergers and acquisitions, along with increased private sector R&D expenditures, continued to grow through the 1990s. According to some industry experts, the acceleration in seed company acquisitions stemmed from efforts by acquiring companies to raise their market share in a market with rich profit potential (Kidd, 1989). Some businesses active in mergers and acquisitions may also have been attempting to consolidate market share and distribution infrastructure/capacity in selected species in anticipation of new biotechnology product developments. In addition, companies may seek to acquire others to facilitate access to protected intellectual property, particularly when licensing is costly (Blonigen and Taylor, 2000).

Table 13. Estimated seed sales and shares of U.S. market for major field crops, 1997

Company	Total sales	Total market share[1]	Corn market share	Soybean market share	Cotton market share
	Million dollars	*Percent*			
Pioneer Hi-Bred	1,178	33.6	42	19	0
Monsanto[2]	541	15.4	14	19	11
Novartis	262	7.5	9	5	0
Delta & Pine Land[3]	79	2.3	0	0	73
Dow Agrosciences/ Mycogen	136	3.9	4	4	0
Golden Harvest	93	2.6	4	0	0
AgrEvo/Cargill	93	2.6	4	0	0
Others	1,121	32.0	23	53	16
Total	3,503	100.0	100	100	100

[1] Total market shares in this table include only corn, soybeans, and cotton.

[2] Monsanto acquired Dekalb in 1997 and Asgrow in 1998.

[3] The merger between Monsanto and Delta & Pine Land in 1998 was called off in December 1999.

Sources: Market shares for corn and soybeans: Hayenga (1998); cotton: USDA, AMS. Total crop sales calculated from acreages and seed cost per acre: USDA's Agricultural Resource Management Survey data (1998) and *Agricultural Statistics* (USDA, 1998).

Some firms evolved toward developing "life sciences" complexes organized around the development of such products as agricultural chemicals, seeds, foods and food ingredients, and pharmaceuticals based on applications of related research in biotechnology and genetics. Monsanto, Novartis, and AgrEvo gained a significant share of the market through such strategic behavior (Begemann, 1997) (table 13). Most of those life sciences companies, however, divested their agricultural operations over the past 3 years (King, 2001; Fulton and Giannakas, 2002).

The changing nature of the seed industry following the entry of large firms has been the subject of much debate. Many large firms enjoy economies of scale in R&D and have been able to subsidize seed research with resources and revenues from other corporate divisions (Butler and Marion, 1985, p. 51). The development of biotechnology has also generated opportunities for economies of scope (i.e., producing several products together at a cost less than producing them separately). According to Fulton and Giannakas (2002), once a specific gene has been isolated (e.g., a gene that confers resistance to a particular herbicide) this gene can be used in a number of crops. Furthermore, the entry of large multinational firms in the industry also expands markets, from domestic or regional to global, increasing sales volume and profits supporting R&D.

The entry of multinational firms in the seed industry may also have drawbacks. First, the relatively small size of the commercial seed market—$5.7 billion in the United States and $25 billion worldwide (table 2)—means that seed divisions in large firms are less likely to exert influence on corporate decisions than those divisions involved in larger markets, such as pharmaceuticals and chemicals (FIS/ASSINSEL, 2000). Second, the time-consuming nature of seed R&D requires a long-term perspective on R&D investments, which may not appeal to a firm's shareholders (Butler and Marion, 1985, p. 51). Third, and most importantly, the presence of large firms in the industry raises concerns about increasing market concentration and oligopolistic competition among and between firms (see, for instance, Leibenluft, 1981; Begemann, 1997; Kalaitzandonakes and Hayenga, 1999).

The seed market is still somewhat small in size, compared with other agricultural input markets, such as the pesticide market (table 14). Still, the total market value of purchased seed in the United States grew substantially in the past three decades. This growth has been particularly rapid in the seed markets for major field crops—corn, soybeans, wheat, and cotton—which constituted 70 percent of the overall seed market in 1982. These markets are dominated by a few large firms which, through strategic corporate behavior, have come to play a central role in some or all of these markets. Together, their seed sales amounted to approximately $4 billion in 1999 (table 14).

Before discussing firm- and crop-specific details of the modern seed industry and its structure, it is useful to review the workings of the seed market, or how seeds are developed, manufactured, and distributed to farmers. Though different types of seed have very distinct production processes and markets, a fairly general description of the process is applicable across all types (see box on the process of seed production).

Table 14. Global seed and pesticide sales of major multinational firms, 1999

Company	Seeds	Pesticides
	Million dollars	
Syngenta (Novartis/AstraZeneca)	1,173	7,030
Aventis (Hoechst & Rhone Poulenc)[1]	135	4,582
Dupont (inc. Pioneer)	1,835	2,309
Monsanto/Pharmacia	600	3,230
Dow Agrosciences	220	2,132

[1] Recently acquired by Bayer.
Source: Merrill Lynch (2000).

THE PROCESS OF SEED PRODUCTION, MARKETING, AND DISTRIBUTION

Different types of seed have very distinct production processes and markets, but a general description of the overall process is applicable across most seed types. The seed firm can be viewed in terms of four separate functions: (1) Plant breeding R&D, (2) seed production, (3) seed conditioning, and (4) seed marketing and distribution.

Plant breeding. Plant breeding constitutes the foundation of the modern seed industry in that it creates a unique and marketable product through the application of science. Plant breeders develop seeds embodying such improvements as high yields, resistance to disease and pests, or traits specific to regional agroclimatic conditions. A seed's success in the market depends primarily on its improved traits, which embody the R&D effort.

The high costs associated with large-scale R&D limit it to a relatively small number of large companies and to Federal Government agencies and land-grant colleges and universities. High R&D costs require private sector varieties to be commercially viable, highly competitive, and well protected by intellectual property rights (IPR). Where each breeder holds the exclusive rights to produce and distribute his or her variety, competition tends to be based more on product performance—yield, disease resistance, quality—than price (Leibenluft, 1981, p. 107). Given the size of their R&D investments, these plant breeders play a central role in managing the entire production, distribution, and marketing processes in the seed industry, resulting in extensive vertical integration of the industry (Butler and Marion, 1985, pp. 18- 19). Moreover, there are economies of scale in R&D, marketing, and distribution, but there are not many economies of scale in seed conditioning (Morris, 1998).

Seed production. Seed firms with a marketable seed product typically contract out the production and multiplication processes to farmers, farmers' associations, and private firms. Breeders provide contract growers the *foundation seed* (parent seed stock produced from the original seed developed by plant breeders) to produce either more foundation seed for continued R&D purposes, or *registered seed* for large-scale production purposes. Registered seed is contracted out in a similar manner to produce *certified seed*, sold to farmers conforming to standards of genetic purity and quality established by State agencies (Agrawal et al., 1998, pp. 104-105, Butler and Marion, 1985, p. 16). The production of certified seed requires strategic planning to ensure that market demand is adequately met. This planning may include determining the quantities of each variety to be produced; determining inventories necessary to produce in excess of forecasted demand to avoid immediate or future shortages; and reducing the risks associated with the unpredictable effects of weather conditions, disease, and pests. Production may also require estimating the quantity of saved seeds farmers will use instead of purchased seed, and the differences in quantity and type of seed demanded in different geographic markets (Leibenluft, 1981, p. 109; Butler and Marion, 1985, pp. 18-19). Corn seed firms, for example, disperse seed-growing contracts throughout the United States (and to countries of the Southern Hemisphere as well) to minimize disease and weather risks and often intentionally overproduce by 25 percent of forecasted demand for the coming season to ensure adequate supply.

The production of both registered and certified seed through contract growers is closely managed by seed firms to ensure that the desirable plant characteristics are carried through to subsequent generations, and to prevent open pollination, disease or pest infestation, or other types of problems that could affect product quality. Contract growers are carefully selected by seed firms and are provided with technical assistance or supervision. Seed firms closely control all stages, from seedbed preparation and planting densities to the timing of input application (Agrawal et al., 1998, pp. 106-107).

Seed conditioning. Once harvested, certified seed is conditioned for sale to farmers, a process that typically includes drying, cleaning, and sorting the seed; treating the seed with insecticides and fungicides; and packaging the seed for distribution and sale (Krull et al., 1998, p. 133; *Seed World*, 1999, p. 41). Seed is also subject to inspection under various State programs to ensure that the final product meets certain quality standards. This inspection may include tests for purity, germination, presence of noxious weed seeds, and moisture content.

Seed marketing and distribution. Large seed firms play a direct role in marketing and distributing their end product to regional, national, and international markets. Many firms also license or outsource marketing and distribution to private firms and individuals to improve access to local markets (Butler and Marion, 1985, 16). Local distribution is typically run by independent agents, such as farmerdealers, farmers' associations, company salespeople, and private wholesalers and retailers. Different distribution channels are used in different regions and markets. In the Midwest, for example, most corn seed is sold to farmers by part-time farmer-dealers who have received training directly from the seed firm. In the South, corn seed sales are channeled through agricultural supply stores. On large farms throughout the country, seed company salespeople sell straight to farmers (Leibenluft, 1981, p. 109).

Seed pricing. The market price of seed incorporates the costs associated with development, production, marketing, and distribution. In the long run, the price must be responsive to the farmers' willingness to purchase while at the same time ensuring a profit margin that provides an attractive return on capital to investors. Furthermore, the price depends on the competitiveness of the particular seed market, and the pricing behavior of those firms that hold large shares of the market.

R&D costs account for an important portion of the market price for seed, particularly for hybrids or transgenic seeds over which private firms own exclusive proprietary rights. In recent decades, private sector R&D costs have been rising with the application of new technologies, and much of the increase in seed prices has been associated with this trend (Krull et al., 1998, pp. 133-134). R&D costs vary among the different seed markets. For example, the corn seed market depends extensively on private sector R&D and passes these costs on to farmers, while the wheat seed market depends largely on public sector research, which is almost cost-free to farmers.

Seed production is another major cost, contributing up to about a quarter of the seed price, but the share of these production costs varies as the marketing and distribution costs change. Production costs include paying farmers to grow seed for exclusive resale to the seed firm. Contract growing typically requires that the seed firm pay a margin above the commodity market price for the seed to ensure that optimal growing conditions are maintained to produce a good quality product (Agrawal et al., 1998, p. 115). For example, for corn, a contract payment formula may be $R=1.1(P_{ch} - 0.08)[2(y\text{-}y_{av})+185]$, where P_{ch} is the expected price of the commodity, such as a futures price, y is the farm corn yield (given a certain nitrogen application and weather) and y_{av} is the regional average yield (also given nitrogen application). Thus, the grower payment is based on an adjusted yield that is equal to a typical yield of 185 bushels per acre plus twice the difference between the farm yield and regional average yields. The grower receives an additional bonus of 10 percent to make the contract desirable (Preckel et al., 1997).

Seed conditioning and treatment may account for around 15 percent of the seed price. This process benefits from scale economies arising from the relatively intensive use of capital equipment. Advertising, promotion, and distribution are other major costs. These costs vary with the stage of the product cycle and their share may account for more than 20 percent of the seed price.

Advertising, promotion, and distribution are other major costs. These costs vary with the stage of the product cycle and their share may account for more than 20 percent of the seed price. Advertising and promotion are necessary to distinguish a seed firm's product from other firms' products on the market, to educate dealers on the best crop management practices to ensure high seed productivity, and to induce farmers to adopt the firm's particular seed.

Distribution costs include costs of transportation and communication between production facility, wholesalers, retailers, and farmers, as well as storage costs (including financial costs) if seed is held as inventory between seasons (Krull et al., 1998, pp. 133-134; Agrawal, 1998, p. 120).

MERGERS AND ACQUISITIONS ROSE IN THE PAST THREE DECADES

In the present seed industry structure, large private firms play a central role in developing and marketing seed for major field crops, such as corn, soybeans, cotton, and wheat. Moreover, the evolution of those firms provides insights into the dynamics of the modern seed industry (see box on evolution of the major seed companies).

In discussing market concentration in the seed industry, it should be noted first that the number of firms participating in the seed industry increases through each step of the production process. Plant breeding is a concentrated stage of the industry, while the production and distribution of certified seed is carried out by hundreds of companies operating in different volumes and markets. A larger numbers of firms are involved in the production and distribution of public varieties: the absence of exclusive property rights means that, in these cases, any individual or firm may produce the seed without permission and may distribute it without licenses (Butler and Marion, 1985, pp. 16-17). Market concentration is usually measured using the four- or eight-firm concentration ratio (CR4 or CR8), which is the share of total industry sales of the four or eight largest firms. Alternatively, market concentration may be measured by the Herfindahl-Hirschmann Index (HHI), which is the sum of squared market shares (in percentage terms) of each firm in the industry. Although it is difficult to precisely determine market size and concentration for the overall seed industry, estimates can be made of individual seed markets for major field crops.

Corn. Since its inception, the corn seed industry has included many small firms—105 of the original 190 companies operating in the 1930s were still in existence in the 1990s—together with larger market leaders, such as Hi-Bred Corn Company (which later became Pioneer), Funk Brothers Seed Company, DeKalb Agricultural Association, and Pfister Hybrid Corn (Duvick, 1998, p. 198). The size and success of the corn seed market is reflected by the fact that nearly all acreage planted in 1997 used seed purchased from the private sector.

Until the 1970s, the corn seed market was characterized by small firms controlling approximately 30 percent of the industry and larger market leaders controlling 70 percent. Between 1973 and 1983, the four largest firms in the U.S. corn seed industry are estimated to have held (CR4) between 50 and 60 percent of the market (figure 11, table 15).

By the mid-1980s, Pioneer had expanded its market share to 38 percent while most other large firms, including Pioneer's largest competitor, DeKalb, experienced sharp declines in their market shares. The decline of other large firms, and the concurrent expansion of market share held by smaller firms, is reflected in a decrease in the CR4 ratio between 1973 and 1983 from 60 to 55 percent. In the 1990s, market concentration in corn seed had grown with the strategic entry into the industry of multinational firms. By 1997, the CR4 ratio had risen to 69 percent, as Pioneer continued to control 42 percent of the market, followed by Monsanto with 14 percent and Novartis/Syngenta with 9 percent (table 16). Smaller firms still control over 20 percent of the market.[9]

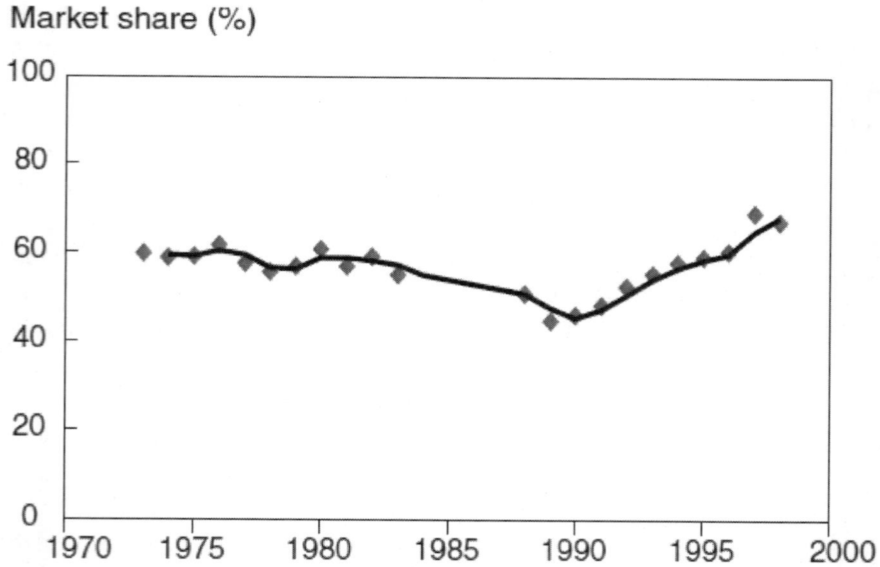

Source: Data sources provided in tables 15 and 16.

Figure 11. Market shares of four largest firms, U.S. corn seed industry.

Table 15. U.S. market shares of corn seed by company[1]

Company	1973	1974	1975	1976	1977	1978	1979	1980	1981	1982	1983
	Percent										
Pioneer	23.8	25.5	24.6	27.3	30.9	26.2	32.9	36.9	34.8	38.8	38.1
DeKalb[2]	21.0	18.8	18.8	19.5	15.8	17.9	13.3	13.0	15.9	12.2	10.3
Asgrow	0	0	0	0	0	0	0	0	0	0	0
Funk[3]	8.8	9.4	8.9	9.2	6.4	8.1	6.7	5.7	5.4	5.2	3.9
Trojan[4]	5.9	5.1	6.8	5.6	4.2	5.4	3.8	2.0	0	0	0
Northrup-King[5]	6.1	4.5	4.7	3.4	3.8	3.3	3.8	4.9	3.4	2.6	2.5
Zeneca/ICI	0	0	0	0	0	0	0	0	0	0	0
Cargill/PAG[6]	4.8	6.8	3.9	3.5	4.1	4.6	3.3	4.7	5.6	5.4	4.2
Golden Harvest	0	0	1.8	2.4	2.5	3.1	2.9	1.3	3.2	2.3	2.6
Dow/Mycogen	0	0	0	0	0	0	0	0	0	0	0
Jacques/Agrigenetics[7]	0	1.3	1.7	2	1.9	2.1	2.7	2.2	0	0	0
Other	29.6	28.6	29.8	27.1	30.4	29.3	30.6	29.3	31.7	33.6	38.4
Largest 8 firms	72.5	70.7	69.8	71.2	68.1	67.0	69.7	69.4	70.0	68.3	64.0
Largest 4 firms	59.7	58.8	59.1	61.6	57.3	55.6	56.7	60.5	59.5	59.1	54.9
Herfindahl index	0.1171	0.1159	0.112	0.1269	0.1049	0.1138	0.1354	0.1609	0.1501	0.1723	0.1604

Note: Due to the sample size of the surveys, the shares are estimates that may vary plus or minus two percentage points.

[1] Market shares are based on percentage of acres sown with respective firm's seed.

[2] Merged with Pfizer in 1982.

[3] Aquired by Ciba-Geigy in 1974.

[4] Aquired by Pfizer in 1975.

[5] Aquired by Sandoz in 1976.

[6] Acquired by Cargill in 1971.

[7] Acquired by Agrigenetics in 1980.

Sources: 1973-80: Butler & Marion (1985), p. 90; 1981-83: McMullen (1987), p. 96.

Table 16. U.S. market shares of corn seed by company

Company	1988	1989	1990	1991	1992	1993	1994	1995	1996	1997	1998
						Percent					
Dupont/											
Pioneer Hi-Bred	34.5	32.0	33.4	35.8	39.6	42.7	44.9	45.0	41.0	42.0	39.0
Monsanto[1]										14.0	15.0
DeKalb	9.0	8.5	9.0	8.0	8.2	8.2	8.6	9.8	10.1	10.0	11.0
Asgrow									2.0	4.0	4.0
Novartis[2]										9.0	9.0
Northrup-King /											
Sandoz[3]	4.0	4.0	3.5	4.4	4.6	4.3	4.2	4.1	5.0		
Dow Agro/ Mycogen[4]									4.3	4.0	4.0
AgrEvo/Cargill[5]	3.0	3.0	3.0	3.5	3.7	3.8	3.9	4.0	3.3	4.0	4.0
Ciba									3.1		
ICI/Zeneca/Advanta[6]									2.9	3.0	3.0
Golden Harvest	2.0	2.0	2.0	2.9	3.0	3.0	3.0	3.3	2.3	4.0	3.0
Others	39.7	42.7	41.7	36.7	33.1	31.3	28.6	28.4	25.6	20.0	23.0[8]
Largest 8 firms[7]	52.5	49.5	50.9	54.6	59.1	62.0	64.6	66.2	72.0	80.0	77.0
Largest 4 firms	50.5	44.5	45.9	48.2	52.4	55.2	57.7	58.9	60.4	69.0	67.0
Herfindahl index	0.1300	0.1125	0.1222	0.1386	0.1679	0.1932	0.2132	0.2165	0.1864	0.2098 0	.1877

[1] Monsanto acquired DeKalb in 1997 and Asgrow in 1998. [2] Result of the merger between Ciba and Sandoz in 1996.

[3] Northrup-King is Sandoz's American seed subsidiary. [4] Mycogen was bought by Dow Agrosciences in 1998.

[5] AgrEvo acquired Cargill's domestic seed business in 1998.

[6] ICI split in 1993 and Zeneca, the pharmaceutical spinoff, was left in control of the company's seed operations. Later in 1996, Zeneca became a part of the Advanta Seed Group. [7] In 1997 and 1998, market shares of only seven companies were available.

[8] Market share amount adjusted from reported figure in Kalaitzandonakes and Hayenga to make the market add up to one.

Source: 1988-95: Merrill Lynch (various years); 1996: Kalaitzandonakes (1997); 1997: Hayenga (1998); 1998: Kalaitzandonakes and Hayenga (1999).

THE EVOLUTION OF THE MAJOR SEED COMPANIES

The evolutionary paths of four major seed companies—Pioneer/Dupont, Novartis/Syngenta, Monsanto, and Advanta Seed Group—provide some insight into the modern structure of the seed industry (figs. A-1-A-4). In each figure, the horizontal arrows pointing to a company indicate an acquisition, vertical arrows pointing down represent a merger, and a line with arrowheads at both ends indicates a joint venture. Where possible, the nationality of the company is given in parentheses next to the company's name, along with the cost and date of acquisition.

Novartis/Syngenta. Novartis was formed in 1996 by the merger of two Swiss life science giants, Ciba-Geigy and Sandoz. Sandoz brought to the merger Northrup-King, a brand name company acquired in 1976 that was well established in field crops, especially hybrid corn and sorghum. Northrup-King's own position in the market was the result of its past acquisitions of field seed companies, including Pride Seed Company, Stauffer Seeds, and Coker Pedigreed Seed. Ciba-Geigy also contributed to the merger with a long list of previously acquired seed companies, including Funk Brothers Seeds and its extensive U.S. distribution network. The 1996 merger gave rise to a new seed division called Novartis Seeds, which controlled 7 percent of the seed market for major crops in 1997. In 1999, after operating as a complete life sciences company for only 3.5 years, Novartis announced plans to merge its agricultural business with the Swedish/English pharmaceutical giant AstraZeneca which had been formed only 6 months earlier. The agricultural spinoff, Syngenta, became a global leader in both seed and pesticide sales. According to the most recent sales figures from Merrill Lynch, Syngenta is only second to Pioneer with $1.2 billion in annual seed sales, and first in pesticide sales with more than $7.0 billion in annual sales (figure A-1).

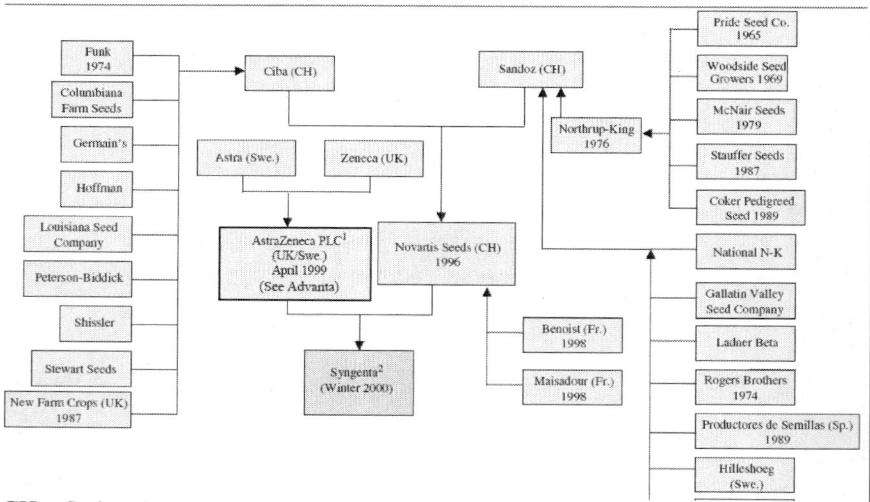

CH = Switzerland.

[1]The merger created the agribusiness spinoff company, Syngenta. The deal did not include Zeneca's stake in Advanta Seed group.

[2]Merger of Novartis's crop protection and seed businesses with AstraZeneca PLC's agrochemicals business.

Sources: Fox, 1990, p. 39-40; Joly and Lemarie, 1999; Leibenluft, 1981, p. 115-116; Northrup-King Co., 2000; Schor, 1984; and Shields, 1999.

Figure A-1. Evolution of Syngenta AG.

Pioneer/Dupont. Pioneer was one of the first four firms active in the emerging corn seed market in the early 1930s. Its modern achievements can be partly attributed to its success as the largest player in the corn seed market for about 40 years (Pioneer, 2001). Between 1973 and 1980, Pioneer made a series of acquisitions that further strengthened its overall position in the seed market: The purchase of Lankhart and Lockett in 1975 allowed Pioneer to expand its activities in the cotton seed (used for planting and not cotton seed sold for oil or other uses) market, while its 1973 purchase of Peterson seeds gave it a larger presence in a soybean market otherwise dominated by public varieties. Bought by the chemicals giant DuPont in 1999, Pioneer continues to operate from its headquarters in Des Moines, Iowa, under the wellestablished Pioneer name as a part of the DuPont conglomerate.

Dupont is the world's leader in production of low-use-rate herbicides, and its acquisition of Pioneer is an important element of its life sciences strategy focusing on the commercialization of a new generation of food, feed, and nutrition products developed with new biotechnology applications (figure A-2)

Monsanto. Barely active in the seed industry until the mid- 1990s, Monsanto, originally considered a chemical, then a pharmaceutical, company, acquired major players in the seed industry in a short period. In 1997, Monsanto bought Asgrow from a Mexican firm, ELM, and Calgene; in 1998, it bought out DeKalb and Cargill's international seed business. Through the acquisition of biotechnology research companies, including Ecogen, Agracetus, and the Plant Breeding Institute, Monsanto also acquired the rights to recently developed seed technologies (figure A-3). Monsanto also attempted to acquire more than 70 percent of the U.S. cotton seed industry with the acquisition of Delta & Pine Land in 1998. Although Monsanto sold its other cotton subsidiary, Stoneville Pedigreed, to make way for the Delta & Pine Land acquisition, the deal was called off in 1999, ultimately leaving Monsanto with no market share in cotton seed. In March 2000, Monsanto merged with Pharmacia & Upjohn, a multinational pharmaceuticals giant. The agricultural side of the merger retained the Monsanto name while the pharmaceutical and related side operates under the name of Pharmacia Corporation. After a partial initial public offering of Monsanto was launched in October 2000, Pharmacia retained 84 percent ownership. Then, on August 13, 2002, Pharmacia Corporation distributed its 84-percent stake in Monsanto Company to Pharmacia shareowners via a special stock dividend. This distribution completed Pharmacia's spinoff of Monsanto and established Monsanto as a 100-percent publicly traded company (PR Newswire, 2002).

In terms of sales, Monsanto is thus the third largest player in the seed industry worldwide and the fourth largest player in the pesticide market (Monsanto, 2000). However, in terms of agricultural biotech products, Monsanto has the largest market share.

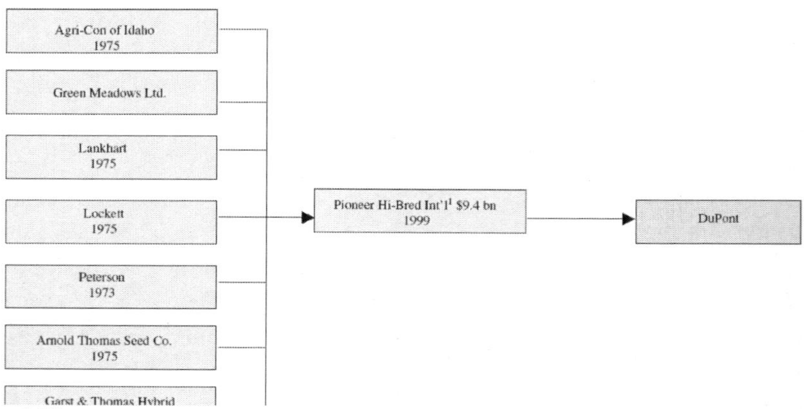

bn = billion.

[1]Dupont bought 20 percent of Pioneer in August 1997 and bought the remaining 80 percent in October 1999. As a DuPont company, Pioneer continues to operate under the Pioneer name from its headquarters in Des Moines, Iowa.

Sources: Butler and Marion, 1985, p. 87; Joly and Lemarie, 1999; Leibenluft, 1981, p. 116; and Seedquest, 1999.

Figure A-2. Evolution of Pioneer/Dupont.

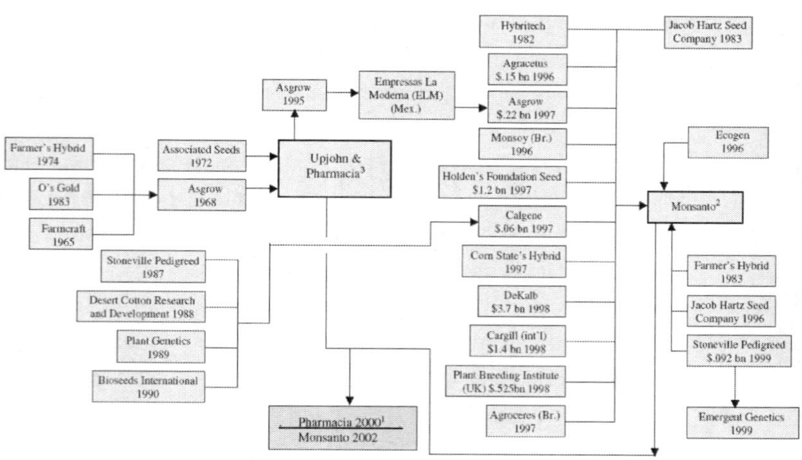

bn = billion. Br. = Brazil.

[1] Monsanto Company became an agricultural subsidiary of Pharmacia Corporation in April 2000. Monsanto became completely separate and independent from Pharmacia on August 13, 2002.

[2] In late December 1999, Monsanto called off its $1.9-bn deal with Delta & Pine Land (D&P). With D&P Land, Monsanto would have acquired more than a 70 percent market share in the cotton seed market since it sold Stoneville Pedigreed in mid-1999 to make way for its acquisition of D&P Land.

[3] Formed in November 1995 by the merger of Pharmacia Aktiebolag and the Upjohn Company, prior to this point, Upjohn had owned Asgrow solely since 1968.

Sources: Asgrow, 2000; Fox, 1999, p. 39; Joly and Lemarie, 1999; Merrill Lynch; Monsanto website; Pharmacia, 2000; PR Newswire, 1999; Schor, 59; Seedquest, 1998; Shimoda, 1999; Wall Street Journal Interactive, 1998.

Figure A-3. Evolution of Monsanto/Asgrow.

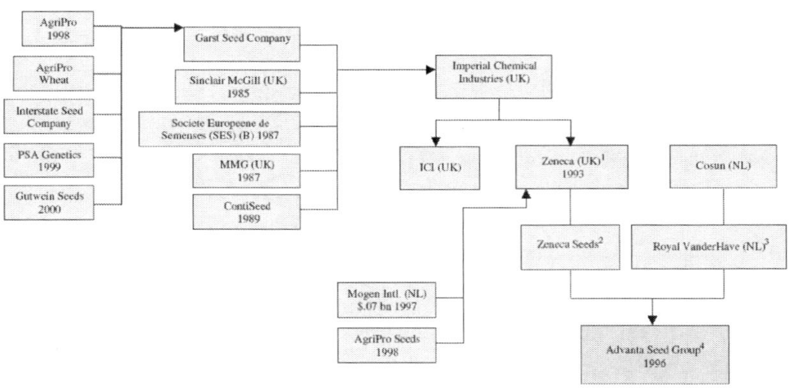

bn = billion. B = Belgium. NL = Netherlands.

[1] On June 1, 1993, ICI split, creating a separately quoted company called Zeneca. Zeneca's industrial pursuits encompass ICI's drugs, agrochemicals and seeds, and specialties businesses.

[2] Zeneca Seeds is a part of Zeneca Agrochemicals. In 1999, Astra (Swe) merged with Zeneca (UK) to form AstraZeneca PLC. The agribusiness merger of Novartis AG and AstraZeneca, called Syngenta, will not change the ownership of Garst Seed Company and its subsidiaries because Advanta was not included in the deal.

[3] Royal VanderHave group is Cosun's international seeds business.

[4] 50:50 joint venture between Zeneca Seeds and Cosun.

Sources: Abrahams, 1993; Clark, 1997; Garst Seed Co., 2000; Zeneca, 2000; Shields, 1998; Fox, 1990, p.40, 85; and Seedquest, 1999.

Figure A-4. Evolution of Advanta Seed Group.

Advanta Seed Group. Advanta Seed Group similarly emerged from numerous acquisitions and joint ventures.

In 1993, Imperial Chemical Industries (ICI), a British chemical manufacturing giant, split its company into ICI and Zeneca. Zeneca took over ICI's production of agrochemicals and seeds. In the deal, the ownership of Garst Seed Company, an American seed firm bought by ICI in 1985, was transferred over to Zeneca, thereby securing Zeneca's position in the U.S. market. In 1996, Zeneca merged its seed business with Royal VanderHave, the international seeds business of Dutch food manufacturer Cosun. Together with Royal VanderHave, Zeneca Seeds forms the backbone of the international seed group Advanta (box figure A-4).

Soybeans. The development of soybean seed varieties was dominated by the public sector until the 1980s (table 17). However, the transformation from public to private was fairly rapid, relative to the transformation in the corn sector. In 1980, over 70 percent of soybean acres harvested in the United States were planted with publicly developed varieties, but by the mid-1990s, the public share had decreased to as low as 10 percent of the market. The increasing role of the private sector appears to have been largely due to the strengthening of intellectual property rights. The private sector's expansion into soybean seed led to a fairly concentrated industry: In the late 1980s, the four largest firms controlled about 40 percent of the soybean seed market, a relatively smaller share than in the corn seed market (table 18) (Knudson and Hansen, 1991).

However, it is difficult to discern a clear leader among these large firms in the soybean seed market. Pioneer may have held a strong position in this market, but with Monsanto's acquisitions of Asgrow and DeKalb, and with the expansion of Novartis into the market, no single firm seems to consistently outsell the others. In fact, figures may indicate that the soybean market is becoming less concentrated over time: The absence of a clear market leader, the presence of a large number of small firms, and a decrease in the HHI between 1994 and 1998 from 0.1115 to 0.0915 all point to decreasing market concentration. Such conclusions, however, depend on the number of small firms categorized as "Others" in tables 18 and 19.

Cotton. Until the early 1980s, private firms and some public institutions maintained a strong presence in the development of cotton seed varieties (table

20). The two largest private firms, Delta & Pine Land and Stoneville, together controlled roughly 40 percent of the varieties planted. Smaller public and private breeders, such as Coker Pedigreed, Lankart, and University of New Mexico AES, each held between 5 and 15 percent.

Table 17. U.S. shares of soybean varieties, public versus private

Year	Public sector varieties	Private sector varieties	Unknown varieties	Varieties from leading four private firms
		Percent of area planted		
980	70	8	22	7
997*	10-30	70-90	--	37-47

-- = not applicable.

* Estimated figures. Smaller figure for public sector (and larger figure for private sector) assumes planted areas are roughly proportional to seed sales. Larger figure for public sector (and smaller figure for private sector) assumes most farmer-saved seed is from public sector varieties. About 25 percent of soybean seed in 1997 was estimated to be farmer saved.

Source: Heisey (1999a).

In the 1980s, the cotton seed market expanded as new developments in cotton breeding brought improved seed varieties to producers, and producers recognized that the traditional practice of cleaning and separating out saved seed was less economical than purchasing seed. Between 1982 and 1997, the use of purchased seed increased from 50 to 75 percent, and large private firms rapidly replaced smaller firms and public institutions as suppliers of seed varieties. Delta & Pine Land continued to lead the market, a position that was strengthened by its acquisition of Paymaster in 1994 and Sure-Grow in 1996, resulting in a 76-percent share of the market by 1999. The second largest firm in the market, Stoneville, controlled only 13 percent of the market in 1999. Of the 19 small breeders (public and private) operating in 1990 with at least a 1-percent share of the market, only 7 still exist. In effect, the cotton market has become highly concentrated, a fact reflected by a CR4 ratio in 1999 of 96 percent (table 20) (figure 12).

Wheat. Most U.S. wheat is cultivated from saved seed, implying that the incentives to private firm entry, research and development, and strategic behavior in the wheat seed industry are fairly limited (table 21). Furthermore, limited development of viable hybrid alternatives to self-pollinated varieties during the 1970s further constrained private sector interest in the market (Fuglie et al., 1996, p. 54-55). See also Knudson (1990) and Hansen and Knudson (1996).

Table 18. U.S. market shares of soybean seed varieties

Institution/Company	1980 Share of acreage harvested with varieties from given breeder	1988 Share of market sales
	Percent	
Major public breeders:		
University of Illinois	20.5	NA
Mississippi AES	16.6	NA
Iowa State University	8.4	NA
University of Florida	6.2	NA
Purdue AES	4.9	NA
Arkansas AES	4.3	NA
Virginia AES	3.4	NA
Minnesota AES	3.2	NA
North Carolina State AES	2.7	NA
Total major public	70.2	30.5
Major private breeders:		
Northrup-King (Sandoz)	2.0	7.6
Asgrow (Upjohn)	1.8	14.9
Pioneer/Peterson	1.4	13.7
Monsanto	0.0	3.4
DeKalb	0.0	5.5
FS	0.0	2.2
Stine	0.0	3.4
North American Plant Breeders (Shell/Olin)	1.4	0.0
Ring Around Products (Occidental Petroleum)	1.0	0.0
Others	22.2	18.8
Total private and public	100.0	100.0
Largest 4 firms	6.6	42.0
Herfindahl index	0.1216	0.0526

NA = not available.

AES = Agricultural Experiment Station.

Sources: 1980: Butler and Marion (1985), p. 91; 1988: Kimle and Hayenga (1992).

Table 19. U.S. market shares of soybean seed, by company

Company	1994	1997	1998
	Percent		
Dupont/Pioneer Hi-Bred[1]	22.0	19.0	17.0
Monsanto[2]		19.0	24.0
Asgrow	15.0	11.0	16.0
DeKalb	19.0	8.0	8.0
Novartis		5.0	5.0
Dow Agrosciences/Mycogen[3]	3.7	4.0	3.0
Stine	4.0	4.0	4.0
FS	3.9		
Jacques			
Others	41.2	39.0	39.0
Public	3.2	10.0	10.0
Largest 8 firms[4]	NA	NA	NA
Largest 4 firms	60.0	47.0	50.0
Herfindahl index[5]	0.1115	0.0779	0.0915

NA = not available.

[1] Pioneer Hi-Bred was fully bought by Dupont in 1999.

[2] Monsanto acquired Asgrow in 1997 and DeKalb in 1998.

[3] Mycogen was acquired by Dow Agro in 1998.

[4] The market shares of only six companies were available in 1994, and only five in 1997/8.

[5] The "others" category was not included in calculation of the Herfindahl index because the category is very large and the number of companies in the others category is also unknown. Public varieties were also not included in this figure.

Sources: 1994: Kalaitzandonakes (1997); Hayenga (1998); 1998: Kalaitzandonakes and Hayenga (1999).

As late as 1997, purchased seed accounted for only 37 percent of all acreage planted with wheat. Moreover, the public sector still plays a central role in the wheat seed market, despite more recent increases in private sector participation (table 22) (Heisey et al., 2001). In 1980, public breeders accounted for 72 percent of sales of hard red spring wheat seed, 80 percent for soft red winter wheat, and 85 percent for hard red winter wheat. Within the public sector, several key institutions were particularly active in providing wheat seed varieties: together the University of Minnesota and University of California at Davis provided 56 percent of the seed varieties for hard red spring wheat in 1980; Purdue University provided 65 percent of the seed varieties for soft red winter wheat; and Kansas State University and University of Nebraska provided a combined 60 percent of the seed varieties for hard red winter wheat. In the private sector, Northrup-King provided 14 percent of the seed varieties for hard red spring wheat in 1980. The total share for the private firms was 18 percent. Coker provided 7 percent of the market for soft red winter wheat, and North American Plant Breeders supplied 5 percent of the market for hard red winter wheat, with smaller firms providing the rest (Butler and Marion, 1985, p. 93).

The Effects of Concentration

The increase in seed industry concentration has raised concerns about its potential impact on market power.[10] However, concentration may result in trade- offs between increased market power and the economies resulting from the increased concentration (arising from mergers or other combinations) (Williamson, 1968).

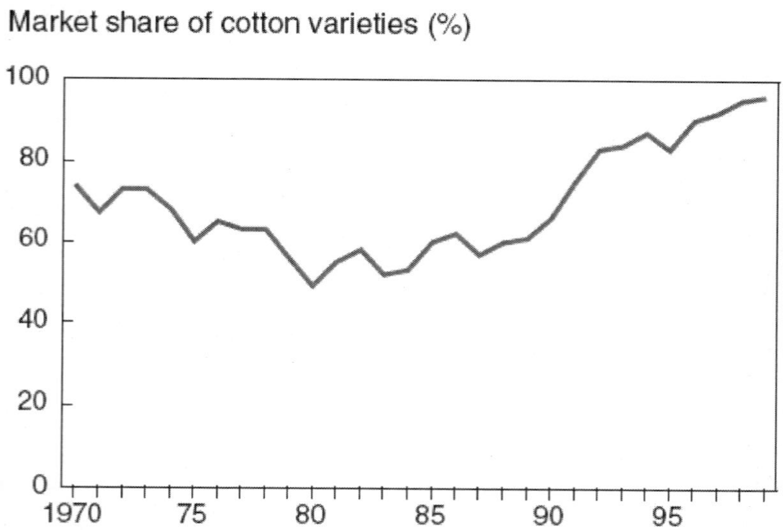

Source: Data source provided in table 20.

Figure 12. Market shares of four largest firms, U.S. cotton seed industry.

Table 20. Share of cotton varieties planted in the United States, by firm or institution

Firm / Institution	1970	1971	1972	1973	1974	1975	1976	1977	1978	1979	1980	1981	1982	1983	1984
	Percent														
Public:															
Arkansas AES (Rex variety)	1	1	2	1	--	--	--	--	--	--	--	*	*	*	*
Mississippi FAES[1] (Dixie King & DES varieties)	2	2	2	1	1	1	1	1	--	--	1	--	1	1	1
Oklahoma AES (Westburn variety)	--	--	--	--		--	--	1	2	1	1	1	1	--	--
Texas AES (Blightmaster & Tamcot varie ties)	1	1	--	2	2	2	6	6	7	8	8	10	9	10	8
University of New Mexico AES (Acala variety)	15	11	11	15	14	10	10	10	13	13	10	13	14	13	11
Private:															
All-Tex	*	*	*	*	*	*	*	*	*	*	*	*	*	*	*
Brownfield Seed & Delinting (Quapaw variety)	*	*	*	*	*	*	1	1	1	2	1	2	1	1	1
California Planting Cotton Seed Distributors (CPCSD)	*	*	*	*	*	*	*	*	*	*	*	*	*	*	*

Table 20. (Continued)

Firm / Institution	1970	1971	1972	1973	1974	1975	1976	1977	1978	1979	1980	1981	1982	1983	1984
	Percent														
Coker Pedigreed Seed	6	8	8	6	8	5	6	4	3	4	3	3	3	3	3
Custom Agricultural Service (Cascot variety)	*	*	*	*	*	*	*	*	1	2	2	2	2	4	3
Delta & Pine Land[2] (DES, Lankart, Paymaster & Sure-Grow varieties)	28	26	29	24	23	17	20	18	18	15	14	16	16	13	16
Dunn Seed Farms	1	1	--	--	2	2	1	1	2	2	2	3	2	4	4
G & P Seed Company	*	*	*	*	*	*	*	*	*	1	1	1	2	1	2
Germain's (Acala variety)	*	*	*	*	*	*	*	*	*	*	*	*	*	*	*
Growers Seed Association (GSA)	*	*	*	*	*	1	3	4	7	7	7	7	2	5	6
Hyperformer	*	*	*	*	*	*	*	*	*	*	*	*	*	*	*
Lankart	10	9	8	10	12	14	12	12	14	13	10	10	10	9	6
Lockett	2	2	2	2	3	4	2	1	2	2	2	2	1	1	--
McNair	1	1	--	1	--	1	1	1	1	1	1	1	2	1	3

Firm / Institution	1970	1971	1972	1973	1974	1975	1976	1977	1978	1979	1980	1981	1982	1983	1984
	Percent														
Monsanto Company[3]	*	*	*	*	*	*	*	*	*	*	*	*	*	*	*
Paymaster	6	5	2	5	6	8	8	9	7	8	7	9	8	9	10
Pioneer (PR 80, Rosebud 80 varieties)	*	*	*	*	*	*	*	*	*	*	*	*	*	1	
Rilcot Seed Company (Stripper & Rilcot varieties)	2	4	3	2	5	6	4	4	2	2	2	1	1	--	--
Stoneville Pedigreed[4] (Stoneville, Coker & McNair varieties)	21	21	25	24	19	19	23	23	18	15	15	16	18	16	16
Summit (Deltapine variety)	*	*	*	*	*	*	*	*	*	*	*	*	*	*	*
Sure-Grow[5]	*	*	*	*	*	*	*	*	*	*	*	*	*	*	*
All other	4	8	8	7	5	10	2	4	2	4	13	3	7	8	10
Largest 8 firms[6]	90	86	88	88	90	83	89	86	87	83	74	84	80	79	77
Largest 4 firms	74	67	73	73	68	60	65	63	63	56	49	55	58	52	53
Herfindahl Index	0.1648	0.1474	0.1778	0.158	0.139	0.1152	0.1345	0.1278	0.1197	0.1014	0.0982	0.1050	.11040	.0992	0.0762
Public:															
Arkansas AES (Rex variety)	*	*	*	*	*	*	*	*	*	*	*	*	*	*	*

Table 20. (Continued)

Firm / Institution	1970	1971	1972	1973	1974	1975	1976	1977	1978	1979	1980	1981	1982	1983	1984
	Percent														
Mississippi AFES[1] (Dixie King & DES varieties)	--	1	1	2	3	3	--	--	--	--	--	--	--	--	--
Oklahoma AES (Westburn variety)	--	--	*	*	*	*	*	*	*	*	*	*	*	*	*
Texas AES (Blightmaster & Tamcot varie ties)	9	7	6	5	7	3	4	1	1	1	1	1	1	1	1
University of New Mexico AES (Acala variety)	10	8	7	5	6	5	3	--	--	--	--	--	--	--	--
Private:															
All-Tex	--	--	1	1	3	1	2	1	1	2	2	3	2	2	2
Brownfield Seed & Delinting (Quapaw variety)	1	1	--	--	*	*	*	*	*	*	*	*	*	*	*
California Planting Cotton Seed Distributors (CPCSD)	*	*	*	*	*	*	*	5	7	6	5	7	6	5	5
Coker Pedigreed Seed	5	3	1	2	1	1	--	--	--	--	--	--	--	--	--
Custom Agri-cultural Service (Cascot variety)	2	2	3	2	1	1	--	1	--	--	--	--	--	--	--

Firm / Institution	1970	1971	1972	1973	1974	1975	1976	1977	1978	1979	1980	1981	1982	1983	1984
	Percent														
Delta & Pine Land[2] (DES, Lankart, Paymaster & Sure-Grow varieties)	23	27	24	30	33	38	43	54	40	71	61	71	73	72	76
Dunn Seed Farms	4	2	2	1	1	1	--	--	--	--	--	--	--	--	--
G & P Seed Company	2	1	2	1	1	1	--	--	--	--	--	--	--	--	--
Germain's (Acala variety)	2	4	4	5	5	5	4	3	1	1	1	1	--	--	--
Growers Seed Association (GSA)	4	4	3	3	1	1	1	*	*	*	*	*	*	*	*
Hyperformer	*	*	*	*	*	*	1	2	2	2	2	2	1	--	--
Lankart	7	8	4	4	4	3	--	--	--	--	--	--	--	--	--
Lockett	--	*	*	*	*	*	*	*	*	*	*	*	*	*	*
McNair	3	2	1	1	1	1	*	*	*	*	*	*	*	*	*
Monsanto Company[3]	*	*	*	*	*	*	*	*	*	*	*	*	11	16	*
Paymaster	9	11	13	14	12	16	22	16	29	**	**	**	**	**	**
Pioneer (PR 80, Rosebud 80 varieties)		1	1	2	1	1	1	1	1	--	1	*	*	*	*
Rilcot Seed Company (Stripper & Rilcot varieties)	*	*	*	*	*	*	*	*	*	*	*	*	*	*	*

Table 20. (Continued)

Firm / Institution	1970	1971	1972	1973	1974	1975	1976	1977	1978	1979	1980	1981	1982	1983	1984
	Percent														
Stoneville Pedigreed[4] (Stoneville, Coker & McNair varieties)	18	16	13	11	9	7	6	8	8	8	9	9	**	**	13
Summit (Delta Pine & Land variety)	*	*	4	3	3	2	*	*	*	*	*	*	*	*	*
Sure-Grow[5]	*	*	*	*	*	--	--	--	1	2	8	**	**	**	**
All other	1	2	10	8	8	10	13	8	9	7	10	6	6	4	3
Largest 8 firms[6]	85	85	74	77	79	80	86	90	89	93	89	95	96	98	100
Largest 4 firms	60	62	57	60	61	66	75	83	84	87	83	90	92	95	96
Herfindahl Index	0.12 44	0.134 2	0.127 2	0.14 1	0.153 8	0.19 2	0.256 4	0.332 8	0.26 28	0.528 8	0.398 4	0.52 12	0.55 1	0.54 76	0.597 8

* indicates that a company left the cotton seed industry or prior to that had presumably not been involved in the industry.

** indicates company was acquired by another company. -- indicates less than a 0.5-percent market share.

AES Stands for Agricultural Experiment Station. These are public research facilities that also release seed varieties.

[1] Mississippi Agricultural and Forestry Experiment Station. [2] Delta Pine & Land acquired Paymaster varieties in 1994.

[3] Monsanto's bid for Delta Pine & Land in 1998 fell through in December 1999 when Monsanto called off the merger.

[4] Stoneville was acquired by Monsanto in 1997 when Monsanto purchased Stoneville's parent company, Calgene. However, Stoneville was then sold by Monsanto in 1999.

[5] Sure-Grow varieties were purchased by Delta & Pine Land in 1996.

[6] For 1994, 1998, and 1999, there were only seven companies listed with more than 0.5-percent share of the cotton seed market.

Sources: USDA, AMS (2000); USDA, ARS (2000).

Table 21. Share of wheat seed sales by principal private and public breeders in the United States, 1980-81

Sector, institution/ company	Hard red spring	Soft red winter	Hard red winter
	Percentage of total sales		
Public sector:			
University of Minnesota	34		
University of California			
(Davis)	22		
North Dakota University	16		
Purdue University		65	
Ohio State University		5	
University of Missouri		5	
University of Arkansas		5	
Kansas State University			35
University of Nebraska			25
Texas A&M University			10
Colorado University			10
Oklahoma State			
University			5
Subtotal	72	80	85
Private sector:			
Northrup-King/McNair (Sandoz)	14	2	
North American Plant			
Breeders (Shell/Olin)		4	5
Pioneer Hi-Bred		3	1
Coker (KWS)		7	
Agrigenetics			1
Western Plant Breeders	1		
World Seeds	2		
Subtotal	18	16	7
Other seeds	10	4	8
Total	100	100	100

Source: Butler and Marion (1985), p. 93.

Table 22. U.S. wheat market shares, public and private varieties, 1981 and 1997

	Unknown	Public sector varieties	Private sector varieties
	Percent of area		*planted*
Hard red winter wheat:			
1981	36	58	6
1997		85	15
Hard red spring wheat:			
1981	37	57	7
1997		85	15
Soft red spring wheat:			
1981	37	63	
1997		35	65

Source:Heisey (1999a).

In the case of the seed industry, if market power dominates, concentration may raise industry profits and margins, and farmers may pay higher-than-competitive prices for seeds. On the other hand, if the efficiency (or cost-reducing) effects outweigh the market power effects, concentration may be beneficial to society.

A growing body of literature presents model-based estimates of the degree of noncompetitive behavior in other industries (Appelbaum, 1979, 1982; Iwata, 1974; Gollop and Roberts, 1979; Azzam and Schroeter, 1995; Azzam, 1997). ERS examined the effects of industry concentration on market power and costs (including R&D) in the U.S. cottonseed and corn seed industries, using an econometric model to measure the relative strengths of these effects over the past 30 years (Fernandez-Cornejo et al., 2002; Fernandez-Cornejo and Spielman, 2002). The model assumes that the profit-maximizing seed firm buys its inputs, including the material input—seed purchased from contract growers—in a competitive market, and sells the seed to cotton (or corn) growers in a noncompetitive market. Most data were collected from USDA and other government sources for the period covering 1970-98. Preliminary empirical results for U.S. cotton and corn seed industries over the past 30 years suggest that increased concentration resulted in a cost-reducing effect that prevailed over the effect of enhanced market power.

ROLES OF PRIVATE AND PUBLIC SECTOR R&D IN CROP SEED HAVE SHIFTED

A prominent change in the seed industry over the last century has been the increasing role of private sector efforts in R&D. Not only have private R&D expenditures increased dramatically in absolute levels, but they have also increased relative to public levels. As private sector R&D expenditures have risen, the types of R&D activity pursued and the choice of crops for research have also changed. With the shift of more R&D activity to the private sector, the process by which the seed industry changes has itself transformed.

Returns to R&D Spending on Plant Breeding

Annual returns to R&D spending on plant breeding exceed 30 percent according to most estimates, although estimates vary widely. Griliches (1958) estimates the returns to public agricultural research for hybrid corn to be 35-40 percent during the period 1940-55. Sundquist et al. (1981) estimate those returns to be 115 percent in 1977. Griliches also finds returns to R&D on hybrid sorghum to be 20 percent during 1940-57. Other studies show returns to cash grains range from 31 to 85 percent (Fuglie et al., pp. 30-31). However, Huffman and Evenson (1993, pp. 245-46) report that returns to public sector crop research (4562 percent) during 1950-82 are lower than returns to private sector research (90 percent) (table 23).

These estimates of returns to R&D investment may fail to incorporate the positive externalities generated by plant breeding research among and between countries. Foreign research in plant breeding benefits from positive spillovers arising from U.S. agricultural research just as U.S. researchers benefit from transfers of genetically diverse materials from research institutions and firms abroad. As a result, consumers in the United States and foreign countries frequently benefit from the increased quality and lower prices offered by new varieties cultivated, imported, or exported in the international economy (Fuglie et al., 1996, p. 28; Fernandez-Cornejo and Shumway, 1997; Maredia and Byerlee, 1999; Schimmelpfennig and Thirtle, 1999; Schimmelpfennig et al., 2000).

Public R&D

Historically in the United States, the public sector has maintained a central role in agricultural R&D. The establishment of the U.S. Department of Agriculture (1862) and the passage of key legislation, such as the Morill Land-Grant College Act (1862), the Hatch Act (1887), and the Smith-Lever Act (1914), expanded this role. The Morill Act established colleges and universities in U.S. States and territories that were dedicated to instruction in agriculture and engineering sciences, and, in 1890, were given further support with Federal funding under the second Morill Act. The Hatch Act provided further support to State-level research by establishing State agricultural experiment stations (SAES) to collaborate with land-grant institutions and to strengthen scientific research in agriculture. The Smith-Lever Act extended this collaboration to include Federal, State, and county agencies through the estab-lishment of the Cooperative Agricultural Extension Service (Fuglie et al., 1996, p. 2). Combined with resources from the USDA and other cooperating government agencies, these legislative acts supported a wide range of public initiatives in agricultural R&D.

Table 23. Estimated returns to crop research in U.S. agriculture, various years

Commodity	Period	Annual return	Study
		Percent	
Hybrid corn	1940-55	35-40	Griliches, 1958
Hybrid sorghum	1940-57	20	Griliches, 1958
Cash grains	1969	47	Bredahl and Peterson, 1976
Crops	1959-64	110	Huffman, 1977
Crops	1964	55	Evenson and Welch, 1979
Cash grains	1969	31-57	Norton, 1981
Cash grains	1974	44-85	Norton, 1981
Maize	1977	115	Sundquist, Cheng, and Norton, 1981
Wheat	1977	97	Sundquist, Cheng, and Norton, 1981
Soybeans	1977	118	Sundquist, Cheng, and Norton, 1981

Commodity	Period	Annual return	Study
		Percent	
Crops - public sector, applied R&D	1950-82	45	Huffman and Evenson, 1993
Crops - public sector, pre-tech R&D	1950-82	62	Huffman and Evenson, 1993
Crops - private sector R&D	1950-82	90	Huffman and Evenson, 1993

Sources: Huffman and Evenson (1993), pp. 245-246; Fuglie et al. (1996), p. 30; Alston and Pardey (1996), pp. 204-206.

One rationale for public investment in agricultural R&D is to address specific market failures. R&D can enhance yields, lower costs, and provide other benefits to both producers and consumers. The incentive for firms to undertake R&D arises from the ability of firms to capture some of the value created from successful innovation. However, the ease of replicating successful R&D undermines the ability of firms to appropriate the returns to their R&D investments (King, 2001). When a competitor can replicate R&D results without incurring the R&D costs, the competitive advantage to firms investing in R&D is not sustainable. The inability of firms to appropriate the returns to their R&D investments results in a market failure, in that productivity-and wealth-enhancing improvements are not attempted.

Other market failures include negative externalities and risk aversion or financial market failures (Beach and Fernandez-Cornejo, 1994). Negative externalities may arise where the social marginal costs of agricultural R&D exceed the private marginal benefits; for instance, when the broad social desire to improve a certain agricultural process is greater than firms' ability to generate such improvements profitably, resulting in underproduction and deadweight losses to society. Risk aversion and financial market failures may occur when private returns from R&D investments over the long term are discounted by investors at a rate higher than the desirable social rate of return.

Modern agricultural R&D includes large amounts of investment from both the public and private sectors (Alston and Pardey, 1996, pp. 29-30). Total public sector expenditure on agricultural R&D, which includes both Federal and State spending, was $3.1 billion in 1996 (table 24). Private sector R&D exceeded $4 billion in the same period (table 25).

Private Sector R&D

The development of commercially viable hybrid corn in the 1930s, the PVPA and subsequent rulings, and other forms of property rights protection led to significant changes in research expenditure patterns and played a key role in the development of new plant varieties. These technological and institutional changes over the past century improved appropriability, increasing incentives for private investment in agricultural R&D, resulting in a larger role in research for private firms.

Real private sector expenditure in agricultural R&D increased by 224 percent from 1960 to 1996 (table 25). Over the same period, real public sector agricultural R&D increased by 97 percent. In addition to the higher relative increase, annual private sector R&D expenditures have exceeded public expenditures every year since 1982. Whereas private sector efforts accounted for slightly less than half of total R&D expenditures from 1960 to 1970, they accounted for 58.7 percent of the total in 1996.

Expenditures on plant breeding and agricultural chemicals were the main areas of increased private sector R&D. From 1960 to 1995, real plant breeding expenditures increased by $514 million (1996 dollars), while R&D on agricultural chemicals increased by $1.392 billion. The growth of R&D in agricultural chemicals primarily reflects the increasing use of herbicides (USDA, 1997, p. 117) and compliance with regulations (Ollinger and Fernandez-Cornejo, 1995).

The dramatic increase in private sector plant breeding R&D expenditures came while public expenditure in that area changed very little in real terms. On the whole, private spending on plant breeding has steadily increased since 1960 as the seed industry increased in size and extent of commercialization. Private sector R&D expenditure has shifted over this period, in percentage terms, from farm machinery and food and kindred products to agricultural chemicals and plant breeding research.[11] These changes in expenditures have been accompanied by structural change in the industry. First, intense merger and acquisition activity in the last three decades led to the formation of large seed conglomerates that allowed once smaller, individually owned, seed companies to take advantage of the strategic R&D relationships and economies of scale of their parent companies. Second, new entrants into the seed industry between 1982 and 1994 increased the number of firms engaged in private plant breeding throughout the United States from 269 to 329 (table 26). This trend in the private sector resulted in a 1,300percent real increase in private R&D plant breeding expenditures between 1960 and 1996.

Table 24. Public agricultural research and development

Year	Biological efficiency expenditure					Non-commodity biotech & biometry expenditure					Pesticides & herbicides expenditure				
	USDA[1]	SAES[1]	Total	Total	Share of public R&D expenditure	USDA[2]	SAES[2]	Total	Total	Share of public R&D expenditure	USDA[3]	SAES[3]	Total	Total	Share of public R&D expenditure
	Million current dollars			*Mil. 1996 dollar*	*Percent*	*Million current dollars*			*Mil.1996 dollar*	*Percent*	*Million current dollars*			*Mil. 1996 dollar*	*Percent*
1960	4.79	29.14	33.93	228.17	14.30	NA	NA	NA	NA	NA	11.60	16.43	28.02	188.43	11.81
1961	5.22	30.99	36.21	236.44	14.32	NA	NA	NA	NA	NA	12.84	17.26	30.10	196.57	11.91
1962	5.69	32.95	38.64	242.75	14.46	NA	NA	NA	NA	NA	13.40	18.96	32.37	203.33	12.11
1963	6.20	35.04	41.24	248.68	14.49	NA	NA	NA	NA	NA	14.42	20.45	34.87	210.27	12.25
1964	6.76	37.27	44.02	255.85	13.67	NA	NA	NA	NA	NA	16.72	23.14	39.86	231.63	12.38
1965	7.37	39.63	47.00	262.28	13.15	NA	NA	NA	NA	NA	19.60	25.82	45.42	253.49	12.71
1966	8.03	42.14	50.17	266.73	12.85	NA	NA	NA	NA	NA	20.25	30.26	50.51	268.53	12.94
1967	8.75	44.82	53.57	269.43	13.13	NA	NA	NA	NA	NA	20.98	32.92	53.90	271.10	13.21
1968	9.77	46.29	56.06	266.00	12.95	NA	NA	NA	NA	NA	20.42	38.44	58.87	279.32	13.60
1969	10.26	49.96	60.21	267.74	13.18	NA	NA	NA	NA	NA	22.17	40.68	62.85	279.46	13.75
1970	9.69	46.55	56.24	232.26	11.41	6.44	19.70	26.14	107.96	5.30	19.95	37.91	57.86	238.96	11.74
1971	9.74	49.81	59.55	232.37	11.19	7.08	20.41	27.48	107.25	5.16	22.35	40.48	62.83	245.18	11.81
1972	12.69	52.16	64.86	241.62	10.34	9.43	22.67	32.10	119.59	5.12	31.12	43.15	74.27	276.69	11.84
1973	13.64	54.46	68.10	239.10	10.15	10.93	25.27	36.19	127.07	5.40	30.68	48.82	79.51	279.14	11.85
1974	13.96	59.66	73.62	238.94	10.10	11.07	26.70	37.77	122.58	5.18	31.85	56.27	88.12	285.97	12.08

Table 24. (Continued)

Year	Biological efficiency expenditure					Non-commodity biotech & biometry expenditure					Pesticides & herbicides expenditure				
	USDA[1]	SAES[1]	Total	Total	Share of public R&D expenditure	USDA[2]	SAES[2]	Total	Total	Share of public R&D expenditure	USDA[3]	SAES[3]	Total	Total	Share of public R&D expenditure
	Million current dollars			*Mil. 1996 dollar*	*Percent*	*Million current dollars*			*Mil.1996 dollar*	*Percent*	*Million current dollars*			*Mil. 1996 dollar*	*Percent*
1975	15.60	68.89	84.50	254.80	10.26	12.35	27.74	40.09	120.90	4.87	33.96	63.67	97.62	294.40	11.85
1976	23.85	88.10	111.95	317.96	12.46	18.90	39.66	58.56	166.31	6.52	51.69	89.93	141.62	402.23	15.76
1977	20.96	89.55	110.51	301.77	10.71	17.96	35.59	53.54	146.21	5.19	47.40	79.40	126.80	346.24	12.29
1978	24.44	96.01	120.45	310.43	10.41	27.78	39.23	67.01	172.70	5.79	56.30	88.29	144.58	372.63	12.50
1979	27.49	101.70	129.19	309.61	10.36	26.71	43.20	69.91	167.55	5.61	62.41	98.60	161.01	385.87	12.91
1980	26.58	117.05	143.62	316.38	10.50	33.69	50.75	84.43	186.00	6.18	65.84	114.72	180.56	397.74	13.21
1981	34.02	126.34	160.35	323.09	10.49	37.91	55.45	93.36	188.10	6.11	75.56	132.95	208.50	420.10	13.64
1982	33.38	135.83	169.20	316.40	10.31	36.59	62.36	98.94	185.02	6.03	76.10	144.74	220.84	412.96	13.45
1983	36.68	138.96	175.64	310.66	10.31	39.84	70.64	110.48	195.40	6.48	79.20	150.41	229.61	406.11	13.48
1984	36.64	142.75	179.40	300.50	10.14	38.52	75.53	114.05	191.04	6.45	76.11	161.24	237.36	397.58	13.42

Year	Biological efficiency expenditure					Non-commodity biotech & biometry expenditure					Pesticides & herbicides expenditure				
	USDA[1]	SAES[1]	Total	Total	Share of public R&D expenditure	USDA[2]	SAES[2]	Total	Total	Share of public R&D expenditure	USDA[3]	SAES[3]	Total	Total	Share of public R&D expenditure
	Million current dollars			*Mil. 1996 dollar*	*Percent*	*Million current dollars*			*Mil.1996 dollar*	*Percent*	*Million current dollars*			*Mil. 1996 dollar*	*Percent*
1985	44.55	153.05	197.61	314.03	10.25	47.72	90.03	137.75	218.90	7.14	88.12	173.12	261.24	415.16	13.55
1986	52.14	163.19	215.33	324.88	10.69	43.75	100.43	144.18	217.54	7.16	87.57	184.78	272.34	410.91	13.52
1987	56.84	164.75	221.59	315.19	10.26	38.29	105.58	143.87	204.65	6.66	94.82	193.81	288.63	410.56	13.36
1988	66.92	168.89	235.81	322.90	10.25	44.01	114.17	158.18	216.59	6.87	103.43	201.83	305.26	418.00	13.27
1989	70.58	180.00	250.58	325.41	10.25	33.81	122.18	155.99	202.58	6.38	105.63	216.56	322.19	418.40	13.17
1990	69.98	194.07	264.05	325.71	10.16	24.12	126.96	151.08	186.36	5.81	115.21	228.54	343.75	424.03	13.23
1991	75.70	199.36	275.06	325.74	9.89	27.26	132.69	159.95	189.42	5.75	131.05	244.63	375.68	444.90	13.51
1992	79.26	201.85	281.11	322.47	9.65	24.35	133.00	157.34	180.50	5.40	134.57	259.73	394.30	452.32	13.54
1993	80.16	205.83	286.00	313.97	9.63	25.39	137.58	162.97	178.91	5.49	135.03	270.47	405.50	445.16	13.65
1994	79.98	210.68	290.66	308.87	9.34	25.63	145.42	171.05	181.77	5.50	138.99	277.58	416.56	442.67	13.39
1995	74.21	211.50	285.71	294.59	9.02	22.71	142.47	165.18	170.32	5.21	143.71	295.96	439.67	453.34	13.87
1996	69.09	198.10	267.20	267.20	8.49	16.14	139.05	155.19	155.19	4.93	141.52	303.91	445.42	445.42	14.15

Table 24. (Continued)

Year	Production mechanization expenditures					Total public agri.R&D expenditures		
	USDA[4]	SAES[4]	Total	Total	Share of public R&D expenditure	Total[5]	Total[5]	Agric.R&D deflator[6]
	Million current dollars			*Mil.1996 dollars*	*Percent*	*Million current dollars*	*Mil.1996 dollars*	
1960	NA	NA	NA	NA	NA	237.3	1,595.7	0.1487
1961	NA	NA	NA	NA	NA	252.8	1,650.6	0.1531
1962	NA	NA	NA	NA	NA	267.3	1,679.3	0.1592
1963	NA	NA	NA	NA	NA	284.7	1,716.6	0.1658
1964	NA	NA	NA	NA	NA	322.0	1,871.5	0.1721
1965	NA	NA	NA	NA	NA	357.3	1,994.4	0.1792
1966	NA	NA	NA	NA	NA	390.5	2,075.9	0.1881
1967	NA	NA	NA	NA	NA	408.1	2,052.4	0.1988
1968	2.37	3.76	6.13	29.09	1.42	432.7	2,053.3	0.2107
1969	2.72	4.33	7.05	31.34	1.54	457.0	2,032.0	0.2249
1970	1.85	4.40	6.24	25.78	1.27	492.8	2,035.3	0.2421
1971	2.97	4.65	7.63	29.76	1.43	532.2	2,076.8	0.2562
1972	3.62	4.47	8.09	30.13	1.29	627.1	2,336.2	0.2684
1973	3.86	4.95	8.81	30.93	1.31	670.7	2,354.9	0.2848
1974	3.93	5.39	9.32	30.26	1.28	729.2	2,366.7	0.3081
1975	4.06	5.67	9.72	29.32	1.18	823.5	2,483.4	0.3316
1976	5.74	7.62	13.36	37.95	1.49	898.4	2,551.6	0.3521
1977	4.82	6.30	11.13	30.38	1.08	1,031.7	2,817.2	0.3662
1978	5.25	7.00	12.25	31.56	1.06	1,157.1	2,982.1	0.3880
1979	5.02	7.27	12.29	29.46	0.99	1,247.2	2,989.1	0.4173

Year	Production mechanization expenditures					Total public agri.R&D expenditures		
	USDA[4]	SAES[4]	Total	Total	Share of public R&D expenditure	Total[5]	Total[5]	Agric.R&D deflator[6]
	Million current dollars			*Mil.1996 dollars*	*Percent*	*Million current dollars*	*Mil.1996 dollars*	
1980	5.48	7.91	13.39	29.49	0.98	1,367.2	3,011.8	0.4540
1981	6.21	7.36	13.57	27.33	0.89	1,528.6	3,079.8	0.4963
1982	5.63	7.97	13.61	25.44	0.83	1,641.6	3,069.6	0.5348
1983	6.22	8.12	14.34	25.36	0.84	1,703.6	3,013.1	0.5654
1984	6.10	7.79	13.89	23.27	0.79	1,769.0	2,963.1	0.5970
1985	6.50	8.96	15.46	24.57	0.80	1,928.0	3,063.9	0.6293
1986	4.32	8.67	12.99	19.60	0.64	2,014.8	3,039.9	0.6628
1987	3.32	9.19	12.51	17.80	0.58	2,160.5	3,073.2	0.7030
1988	3.22	9.77	12.99	17.78	0.56	2,301.2	3,151.0	0.7303
1989	2.67	10.37	13.04	16.93	0.53	2,445.8	3,176.2	0.7700
1990	3.08	10.26	13.35	16.46	0.51	2,598.3	3,205.1	0.8107
1991	3.40	9.56	12.96	15.35	0.47	2,780.5	3,292.8	0.8444
1992	3.24	8.99	12.23	14.03	0.42	2,913.2	3,341.8	0.8717
1993	3.36	9.32	12.68	13.92	0.43	2,970.9	3,261.5	0.9109
1994	3.46	9.98	13.44	14.28	0.43	3,111.5	3,306.6	0.9410
1995	3.24	9.64	12.88	13.28	0.41	3,168.8	3,267.3	0.9698
1996	2.88	9.03	11.91	11.91	0.38	3,148.0	3,148.0	1.0000

NA = not available. SAES = State Agricultural Experiment Station.

[1] Research problem area code 304 and 307 in CRIS (USDA, CSREES, CRIS, 1993). Source: "Inventory of Agricultural Research" for fiscal years 1968-96 (USDA, CSREES, CRIS, various years). For 1960-67, total public expenditures on biological efficiency and pesticides/pest management estimated by linear interpolation.

[2] Research problem area code 318 in CRIS (USDA, CSREES, CRIS, 1993). Source: "Inventory of Agricultural Research" for fiscal years 1968-96 (USDA, CSREES, CRIS, various years). For 1960-67, total public expenditures on biological efficiency and pesticides/pest management are estimated by linear interpolation.

[3] Research problem area codes 204-209 in CRIS (USDA, CSREES, CRIS, 1993). Source: 1960-67: USDA and SAES expenditure on pesticides and herbicides are linear interpolations of data based on total SAES R&D expenditure derived from the rate of growth of total SAES R&D expenditure assumed to be consistent with the annual rate of growth in Alston and Pardey (1996, Table 2-A3, 76); 1968-96:"Inventory of Agricultural Research" for fiscal years 1968-96 (USDA, CSREES, CRIS, various years).

[4] Research problem area code 305 and 308 in CRIS (USDA, CSREES, CRIS, 1993). Source: "Inventory of Agricultural Research" for fiscalyears 1968-96 (USDA, CSREES, CRIS, various years). For 1960-67, total public expenditures on biological efficiency and pesticides/pest management are estimated by linear interpolation.

[5] Total agri. R&D expenditures are not the sum of the four categories of R&D presented, and includes other categories not shown here.Source: 1960-69: data are based on rates of change from Alston and Pardey (1986, p. 76), and Huffman & Evenson (1993, pp. 95-96);1970-96: "Inventory of Agricultural Research" for fiscal years 1970-96 (USDA, CSREES, CRIS, various years).

[6] Source: Klotz-Ingram (2000).

Table 25. Private agricultural research and development

	Plant breeding expenditure			Agricultural chemicals expenditure			Farm machinery expenditure			Food & kindred Products expenditure			Veterinary pharmaceuticals expenditure			Total private agric. R&D expenditure		
	Total[1]	Total	Share of private R&D	Total[1]	Total	Share of private R&D expenditure	Total[1]	Total	Share of private R&D expenditure	Total[1]	Total	Share of private R&D expenditure	Total[1]	Total	Percent of private R&D expenditure	Total[1]	Total	Agric. R&D deflator[2]
	Million current $	*Mil. 1996 dollars*	*Percent*	*Million current $*	*Mil. 1996 dollars*	*Percent*	*Million current $*	*Mil. 1996 dollars*	*Percent*	*Million current $*	*Mil. 1996 dollars*	*Percent*	*Million current $*	*Mil. 1996 dollars*	*Percent*	*Million current $*	*Mil. 1996 dollars*	
1960	6.00	40.35	2.91	27.00	181.56	13.11	75.00	504.34	36.41	92.00	618.66	44.66	6.00	40.35	2.91	206.00	1,385.26	0.1487
1961	6.00	39.18	2.83	38.00	248.14	17.92	65.00	424.45	30.66	92.00	600.76	43.40	11.00	71.83	5.19	212.00	1,384.36	0.1531
1962	6.00	37.69	2.61	42.00	263.84	18.26	70.00	439.73	30.43	98.00	615.62	42.61	14.00	87.95	6.09	230.00	1,444.82	0.1592
1963	7.00	42.21	2.86	45.00	271.33	18.37	76.00	458.25	31.02	102.00	615.02	41.63	15.00	90.44	6.12	245.00	1,477.24	0.1658
1964	8.00	46.49	2.93	48.00	278.96	17.58	79.00	459.12	28.94	118.00	685.77	43.22	20.00	116.23	7.33	273.00	1,586.57	0.1721

Table 25. (Continued)

	Plant breeding expenditure			Agricultural chemicals expenditure			Farm machinery expenditure			Food & kindred Products expenditure			Veterinary pharmaceuticals expenditure			Total private agric. R&D expenditure		
	Total[1]	Total	Share of private R&D	Total[1]	Total	Share of private R&D expenditure	Total[1]	Total	Share of private R&D expenditure	Total[1]	Total	Share of private R&D expenditure	Total[1]	Total	Percent of private R&D expenditure	Total[1]	Total	Agric. R&D deflator[2]
	Million current $	*Mil. 1996 dollars*	*Percent*	*Million current $*	*Mil. 1996 dollars*	*Percent*	*Million current $*	*Mil. 1996 dollars*	*Percent*	*Million current $*	*Mil. 1996 dollars*	*Percent*	*Million current $*	*Mil. 1996 dollars*	*Percent*	*Million current $*	*Mil. 1996 dollars*	
1965	9.00	50.23	2.79	64.00	357.19	19.81	96.00	535.78	29.72	131.00	731.12	40.56	23.00	128.36	7.12	323.00	1,802.68	0.1792
1966	11.00	58.48	3.18	77.00	409.36	22.25	100.00	531.63	28.90	130.00	691.13	37.57	28.00	148.86	8.09	346.00	1,839.46	0.1881
1967	12.00	60.36	3.38	72.00	362.14	20.28	102.00	513.04	28.73	134.00	673.99	37.75	35.00	176.04	9.86	355.00	1,785.57	0.1988
1968	17.00	80.67	4.34	78.00	370.12	19.90	96.00	455.53	24.49	165.00	782.94	42.09	36.00	170.82	9.18	392.00	1,860.07	0.2107
1969	22.00	97.82	5.21	85.00	377.96	20.14	99.00	440.21	23.46	182.00	809.27	43.13	34.00	151.18	8.06	422.00	1,876.44	0.2249

Table 25. Private agricultural research and development

	Plant breeding expenditure			Agricultural chemicals expenditure			Farm machinery expenditure			Food & kindred Products expenditure			Veterinary pharmaceuticals expenditure			Total private agric. R&D expenditure		
	Total[1]	Total	Share of private R&D	Total[1]	Total	Share of private R&D expenditure	Total[1]	Total	Share of private R&D expenditure	Total[1]	Total	Share of private R&D expenditure	Total[1]	Total	Percent of private R&D expenditure	Total[1]	Total	Agric. R&D deflator[2]
	Million current $	*Mil. 1996 dollars*	*Percent*	*Million current $*	*Mil. 1996 dollars*	*Percent*	*Million current $*	*Mil. 1996 dollars*	*Percent*	*Million current $*	*Mil. 1996 dollars*	*Percent*	*Million current $*	*Mil. 1996 dollars*	*Percent*	*Million current $*	*Mil. 1996 dollars*	
1960	6.00	40.35	2.91	27.00	181.56	13.11	75.00	504.34	36.41	92.00	618.66	44.66	6.00	40.35	2.91	206.00	1,385.26	0.1487
1961	6.00	39.18	2.83	38.00	248.14	17.92	65.00	424.45	30.66	92.00	600.76	43.40	11.00	71.83	5.19	212.00	1,384.36	0.1531
1962	6.00	37.69	2.61	42.00	263.84	18.26	70.00	439.73	30.43	98.00	615.62	42.61	14.00	87.95	6.09	230.00	1,444.82	0.1592
1963	7.00	42.21	2.86	45.00	271.33	18.37	76.00	458.25	31.02	102.00	615.02	41.63	15.00	90.44	6.12	245.00	1,477.24	0.1658
1964	8.00	46.49	2.93	48.00	278.96	17.58	79.00	459.12	28.94	118.00	685.77	43.22	20.00	116.23	7.33	273.00	1,586.57	0.1721

Table 25. (Continued)

	Plant breeding expenditure			Agricultural chemicals expenditure			Farm machinery expenditure			Food & kindred Products expenditure			Veterinary pharmaceuticals expenditure			Total private agric. R&D expenditure		
	Total[1]	Total	Share of private R&D	Total[1]	Total	Share of private R&D expenditure	Total[1]	Total	Share of private R&D expenditure	Total[1]	Total	Share of private R&D expenditure	Total[1]	Total	Percent of private R&D expenditure	Total[1]	Total	Agric. R&D deflator[2]
	Million current $	*Mil. 1996 dollars*	*Percent*	*Million current $*	*Mil. 1996 dollars*	*Percent*	*Million current $*	*Mil. 1996 dollars*	*Percent*	*Million current $*	*Mil. 1996 dollars*	*Percent*	*Million current $*	*Mil. 1996 dollars*	*Percent*	*Million current $*	*Mil. 1996 dollars*	
1965	9.00	50.23	2.79	64.00	357.19	19.81	96.00	535.78	29.72	131.00	731.12	40.56	23.00	128.36	7.12	323.00	1,802.68	0.1792
1966	11.00	58.48	3.18	77.00	409.36	22.25	100.00	531.63	28.90	130.00	691.13	37.57	28.00	148.86	8.09	346.00	1,839.46	0.1881
1967	12.00	60.36	3.38	72.00	362.14	20.28	102.00	513.04	28.73	134.00	673.99	37.75	35.00	176.04	9.86	355.00	1,785.57	0.1988
1968	17.00	80.67	4.34	78.00	370.12	19.90	96.00	455.53	24.49	165.00	782.94	42.09	36.00	170.82	9.18	392.00	1,860.07	0.2107
1969	22.00	97.82	5.21	85.00	377.96	20.14	99.00	440.21	23.46	182.00	809.27	43.13	34.00	151.18	8.06	422.00	1,876.44	0.2249

	Plant breeding expenditure			Agricultural chemicals expenditure			Farm machinery expenditure			Food & kindred Products expenditure			Veterinary pharmaceuticals expenditure			Total private agric. R&D expenditure		
	Total[1]	Total	Share of private R&D	Total[1]	Total	Share of private R&D expenditure	Total[1]	Total	Share of private R&D expenditure	Total[1]	Total	Share of private R&D expenditure	Total[1]	Total	Percent of private R&D expenditure	Total[1]	Total	Agric. R&D deflator[2]
	Million current $	*Mil. 1996 dollars*	*Percent*	*Million current $*	*Mil. 1996 dollars*	*Percent*	*Million current $*	*Mil. 1996 dollars*	*Percent*	*Million current $*	*Mil. 1996 dollars*	*Percent*	*Million current $*	*Mil. 1996 dollars*	*Percent*	*Million current $*	*Mil. 1996 dollars*	
1970	26.00	107.38	5.60	98.00	404.72	21.12	89.00	367.55	19.18	206.00	850.74	44.40	45.00	185.84	9.70	464.00	1,916.23	0.2421
1971	29.00	113.17	5.95	109.00	425.37	22.38	90.00	351.22	18.48	211.00	823.42	43.33	48.00	187.32	9.86	487.00	1,900.49	0.2562
1972	32.00	119.21	6.30	104.00	387.44	20.47	93.00	346.46	18.31	227.00	845.67	44.69	52.00	193.72	10.24	508.00	1,892.51	0.2684
1973	39.00	136.92	6.76	113.00	396.73	19.58	120.00	421.31	20.80	243.00	853.14	42.11	62.00	217.67	10.75	577.00	2,025.78	0.2848
1974	45.00	146.05	6.73	136.00	441.38	20.33	131.00	425.16	19.58	283.00	918.47	42.30	74.00	240.16	11.06	669.00	2,171.22	0.3081
1975	50.00	150.78	7.05	169.00	509.64	23.84	138.00	416.15	19.46	273.00	823.26	38.50	79.00	238.23	11.14	709.00	2,138.06	0.3316

Table 25. (Continued)

	Plant breeding expenditure			Agricultural chemicals expenditure			Farm machinery expenditure			Food & kindred Products expenditure			Veterinary pharmaceuticals expenditure			Total private agric. R&D expenditure		
	Total[1]	Total	Share of private R&D	Total[1]	Total	Share of private R&D expenditure	Total[1]	Total	Share of private R&D expenditure	Total[1]	Total	Share of private R&D expenditure	Total[1]	Total	Percent of private R&D expenditure	Total[1]	Total	Agric. R&D deflator[2]
	Million current $	*Mil. 1996 dollars*	*Percent*	*Million current $*	*Mil. 1996 dollars*	*Percent*	*Million current $*	*Mil. 1996 dollars*	*Percent*	*Million current $*	*Mil. 1996 dollars*	*Percent*	*Million current $*	*Mil. 1996 dollars*	*Percent*	*Million current $*	*Mil. 1996 dollars*	
1976	55.00	156.21	6.72	200.00	568.05	24.45	168.00	477.16	20.54	308.00	874.80	37.65	87.00	247.10	10.64	818.00	2,323.32	0.3521
1977	58.00	158.38	6.08	243.00	663.55	25.47	221.00	603.47	23.17	348.00	950.27	36.48	84.00	229.37	8.81	954.00	2,605.04	0.3662
1978	69.00	177.83	6.39	290.00	747.40	26.88	249.00	641.73	23.08	385.00	992.24	35.68	86.00	221.64	7.97	1,079.00	2,780.85	0.3880
1979	81.00	194.13	6.73	312.00	747.74	25.91	295.00	707.00	24.50	420.00	1,006.58	34.88	96.00	230.08	7.97	1,204.00	2,885.53	0.4173
1980	97.00	213.68	6.67	395.00	870.14	27.17	363.00	799.65	24.97	488.00	1,075.01	33.56	111.00	244.52	7.63	1,454.00	3,202.99	0.4540

	Plant breeding expenditure			Agricultural chemicals expenditure			Farm machinery expenditure			Food & kindred Products expenditure			Veterinary pharmaceuticals expenditure			Total private agric. R&D expenditure		
	Total[1]	Total	Share of private R&D	Total[1]	Total	Share of private R&D expenditure	Total[1]	Total	Share of private R&D expenditure	Total[1]	Total	Share of private R&D expenditure	Total[1]	Total	Percent of private R&D expenditure	Total[1]	Total	Agric. R&D deflator[2]
	Million current $	*Mil. 1996 dollars*	*Percent*	*Million current $*	*Mil. 1996 dollars*	*Percent*	*Million current $*	*Mil. 1996 dollars*	*Percent*	*Million current $*	*Mil. 1996 dollars*	*Percent*	*Million current $*	*Mil. 1996 dollars*	*Percent*	*Million current $*	*Mil. 1996 dollars*	
1981	105.00	211.56	7.15	469.00	944.95	31.93	278.00	560.12	18.92	492.00	991.29	33.49	125.00	251.85	8.51	1,469.00	2,959.78	0.4963
1982	118.00	220.65	7.15	527.00	985.45	31.92	281.00	525.45	17.02	596.00	1,114.47	36.10	129.00	241.22	7.81	1,651.00	3,087.24	0.5348
1983	138.00	244.08	7.69	584.00	1,032.93	32.53	290.00	512.93	16.16	636.00	1,124.90	35.43	147.00	260.00	8.19	1,795.00	3,174.85	0.5654
1984	154.00	257.96	7.53	624.00	1,045.22	30.50	311.00	520.94	15.20	803.00	1,345.05	39.25	154.00	257.96	7.53	2,046.00	3,427.13	0.5970
1985	179.00	284.47	8.26	683.00	1,085.42	31.52	304.00	483.11	14.03	842.00	1,338.10	38.86	159.00	252.68	7.34	2,167.00	3,443.78	0.6293
1986	204.00	307.79	8.79	691.00	1,042.57	29.77	307.00	463.20	13.23	940.00	1,418.25	40.50	179.00	270.07	7.71	2,321.00	3,501.88	0.6628

Table 25. (Continued)

	Plant breeding expenditure			Agricultural chemicals expenditure			Farm machinery expenditure			Food & kindred Products expenditure			Veterinary pharmaceuticals expenditure			Total private agric. R&D expenditure		
	Total[1]	Total	Share of private R&D	Total[1]	Total	Share of private R&D	Total[1]	Total	Share of private R&D expenditure	Total[1]	Total	Share of private R&D expenditure	Total[1]	Total	Percent of private R&D expenditure	Total[1]	Total	Agric. R&D deflator[2]
	Million current $	*Mil. 1996 dollars*	*Percent*	*Million current $*	*Mil. 1996 dollars*	*Percent*	*Million current $*	*Mil. 1996 dollars*	*Percent*	*Million current $*	*Mil. 1996 dollars*	*Percent*	*Million current $*	*Mil. 1996 dollars*	*Percent*	*Million current $*	*Mil. 1996 dollars*	
1987	222.00	315.78	9.75	682.00	970.10	29.94	292.00	415.35	12.82	891.00	1,267.39	39.11	191.00	271.69	8.38	2,278.00	3,240.31	0.7030
1988	245.00	335.48	9.53	938.00	1,284.41	36.48	295.00	403.95	11.47	871.00	1,192.67	33.88	221.00	302.62	8.60	2,571.00	3,520.50	0.7303
1989	283.00	367.51	10.31	979.00	1,271.36	35.65	320.00	415.56	11.65	921.00	1,196.04	33.54	243.00	315.57	8.85	2,746.00	3,566.05	0.7700
1990	314.00	387.33	10.57	1,127.00	1,390.21	37.93	360.00	444.08	12.12	925.00	1,141.03	31.13	245.00	302.22	8.25	2,971.00	3,664.86	0.8107
1991	342.00	405.01	10.78	1,228.00	1,454.25	38.69	382.00	452.38	12.04	946.00	1,120.30	29.80	276.00	326.85	8.70	3,174.00	3,758.80	0.8444

	Plant breeding expenditure			Agricultural chemicals expenditure			Farm machinery expenditure			Food & kindred Products expenditure			Veterinary pharmaceuticals expenditure			Total private agric. R&D expenditure		
	Total[1]	Total	Share of private R&D	Total[1]	Total	Share of private R&D expenditure	Total[1]	Total	Share of private R&D expenditure	Total[1]	Total	Share of private R&D expenditure	Total[1]	Total	Percent of private R&D expenditure	Total[1]	Total	Agric. R&D deflator[2]
	Million current $	*Mil. 1996 dollars*	*Percent*	*Million current $*	*Mil. 1996 dollars*	*Percent*	*Million current $*	*Mil. 1996 dollars*	*Percent*	*Million current $*	*Mil. 1996 dollars*	*Percent*	*Million current $*	*Mil. 1996 dollars*	*Percent*	*Million current $*	*Mil. 1996 dollars*	
1992	400.00	458.86	12.47	1,062.00	1,218.28	33.10	394.00	451.98	12.28	1021.00	1,171.24	31.83	331.00	379.71	10.32	3,208.00	3,680.07	0.8717
1993	453.00	497.31	13.04	1,389.00	1,524.86	39.99	369.00	405.09	10.62	995.00	1,092.32	28.65	267.00	293.12	7.69	3,473.00	3,812.71	0.9109
1994	470.00	499.46	13.10	1,356.00	1,440.99	37.80	377.00	400.63	10.51	1084.00	1,151.94	30.22	300.00	318.80	8.36	3,587.00	3,811.82	0.9410
1995	524.00	540.29	13.64	1,419.00	1,463.12	36.94	436.00	449.56	11.35	1146.00	1,181.63	29.84	316.00	325.82	8.23	3,841.00	3,960.41	0.9698
1996	554.00	554.00	12.35	NA	NA	NA	487.00	487.00	10.86	NA	NA	NA	359.00	359.00	8.00	4,486.00	4,486.00	1.0000

NA = not available.

[1] Source: 1960-92: Klotz, Fuglie, and Pray (1995, p. 26);1993-1997: Klotz, Fuglie and Pray (1995, updated July 1998). [2] Klotz-Ingram, Private Communication (2000).

As the emphasis of R&D shifted, the share of public sector R&D expenditures on plant breeding research remained relatively unchanged at about 10 percent between 1970 and 1990, but the share declined in the 1990s (Figure 13). Public spending on biological efficiency (used as a proxy for public spending on plant breeding) decreased as a share of total public agricultural R&D expenditures, reaching 9 percent ($291 million) of total public sector agricultural R&D in 1994 (figures 13-14).[12] This decrease occurred despite evidence suggesting that the rate of return on public research remains positive, and that such areas as pre-commercial agricultural research continue to require government support (Fuglie et al., 1996, pp. 29-31). On the other hand, the share of private sector research spent on plant breeding increased, reaching 13 percent ($470 million) of total expenditures on private agricultural R&D in 1994 (table 25) (Klotz et al.).

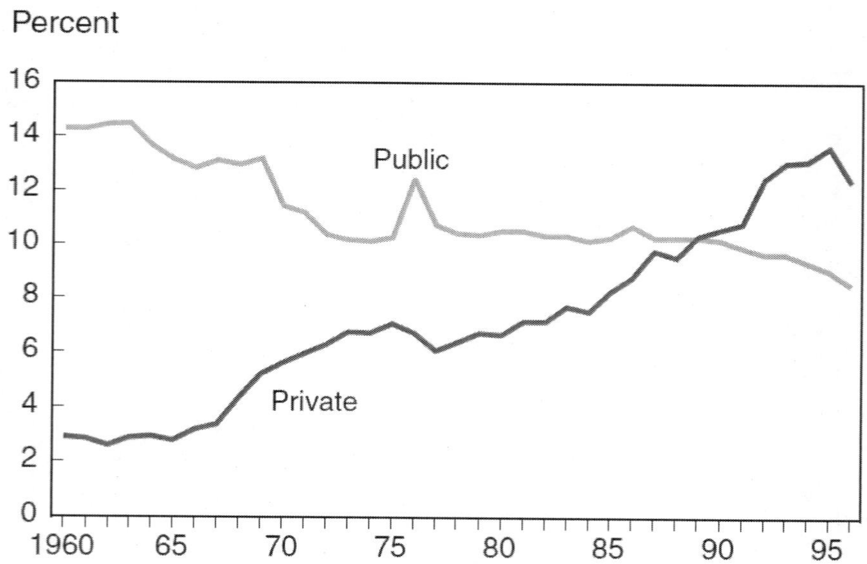

Source: Data source provided in tables 24 and 25.

Figure 13. Plant breeding as a share of total agricultural R&D expenditures.

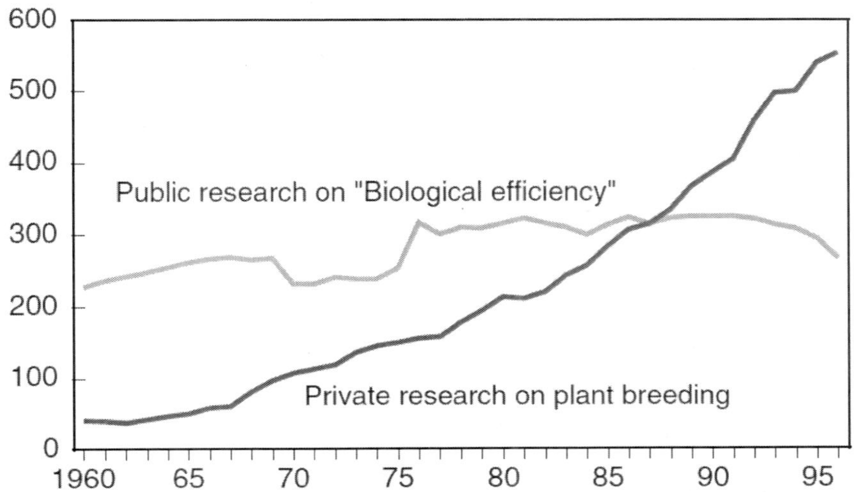

"Biological efficiency" includes breeding and selection of improved plant varieties.
Source: Data source provided in tables 24 and 25.

Figure 14. Public and private research expenditures on plant breeding.

Table 26. Private sector firms engaged in plant breeding, major field crops

Crop	1982		1989		1994	
	Number of companies	**Share of companies**	**Number of companies**	**Share of companies**	**Number of companies**	**Share of companies**
		Percent		*Percent*		*Percent*
Corn	66	24.5	75	27.6	91	27.7
Soybeans	26	9.7	34	12.5	38	11.6
Cotton	13	4.8	11	4.0	35	10.6
Wheat	21	7.8	11	4.0	27	8.2
Others	143	53.2	141	51.8	138	41.9
Total	269	100.0	272	100.0	329	100.0

Source: Companies and expenditures for 1994: Frey (1996, p.19); companies and expenditures for 1982 and 1989:Kalton et al.(1990, p.24).

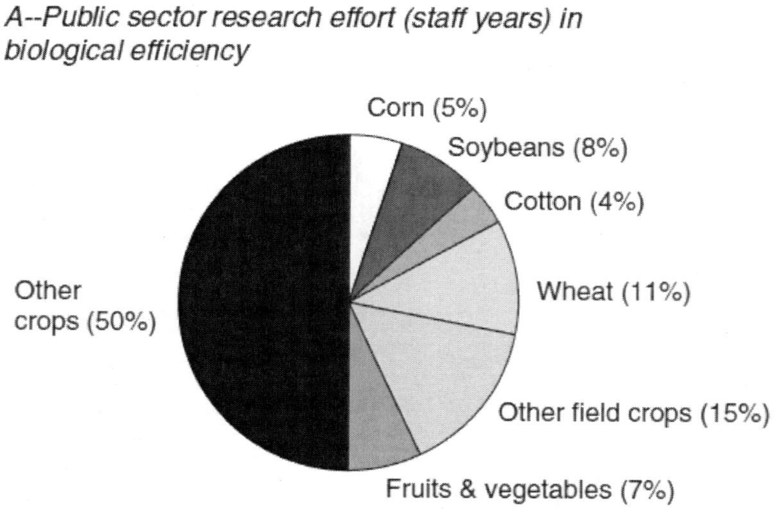

Source: Calculated from the number of SY given in table 32 first column.

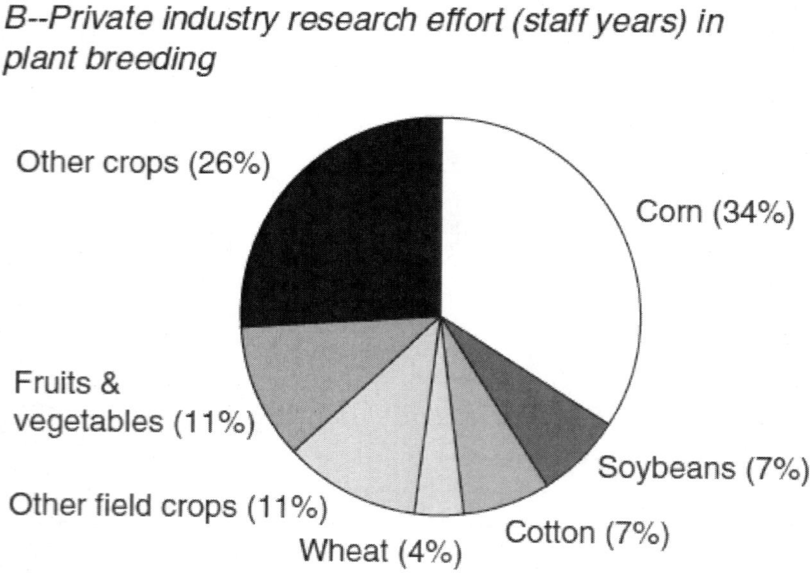

Source: Data source provided in table 31.

Figure 15. Research effort by crop, 1994.

Private and public research do not always emphasize the same areas. The emphasis of the private sector on pure line field crops (Figure 15, table 26) suggests that one role for the public sector is to carry out research in otherwise neglected crops. Private sector research has expanded to include cultivar development on hybrid crops and pre-breeding activities; meanwhile, public plant breeding research has focused on basic germplasm and applied plant genetics. (Heisey et al., 2001). So even as private firms engage in R&D that once was performed mostly by the public sector, important roles still exist for both public and private R&D.

A breakdown of expenditures on plant breeding R&D by specific private firms provides additional insight into the magnitude and growth of private sector research (table 27). Large firms, such as ICI, Sandoz, and Pioneer, each spent between \$38 million and \$57 million on plant breeding in 1988. As a share of seed sales, these figures range from 6.3 percent (Pioneer) to 23.2 percent (ICI). The seed industry considers 5 to 7 percent of sales to be the minimum requisite investment in R&D to maintain competitiveness (James, 1997, p. 6). Other large firms, such as Ciba-Geigy, DeKalb, Limagrain, KWS, and Upjohn, spent between \$16 million and \$27 million each on plant breeding R&D in 1988, and, when measured as a share of sales, each of these firms also exceeded the minimum requisite investment levels in R&D. By 1996, after new rounds of mergers and acquisitions, the upper bounds of R&D expenditures by large firms had increased significantly: Pioneer spent \$133 million on seed R&D, a 2.5-fold increase relative to 1988, followed closely by Novartis with \$122 million. Even Cargill, a smaller player in the market in terms of annual seed revenue, spent \$37 million on seed R&D in 1996. As a share of sales, these expenditure levels again exceeded the industry estimates of minimum spending necessary to sustain competitiveness in the seed market.

Plant Breeding Research Patterns by Crop

The changing focus of public and private sector R&D expenditures in plant breeding has also been associated with changes in research expenditure on specific crops. Historically, public sector expenditures on plant breeding have been allocated to basic and applied research on new varieties of field crops, while private sector expenditures have focused on the development of new plant varieties for home garden and horticultural crops (Fuglie et al., 1996, p. 53).

Table 27. Plant breeding R&D expenditures by company

Company	1986 Plant breeding			1988-89 Plant breeding					1996 Seed		
	Conventional breeding R&D	Plant biotechnology R&D	Total plant breeding R&D	Conventional breeding R&D	Plant biotechnology R&D	Total plant breeding R&D	Seed sales (fiscal 1989)	R&D as share of seed sales	Seed R&D	Annual revenue	R&D as share of revenue
		Million current dollars			*Million current dollars*			*Percent*	*Million current dollars*		*Percent*
Aventis Cropscience											
Rhone-Poulenc AgrEvo	2	3	5				125				
Novartis Zeneca									122	970	12.6
Advanta									53	470	11.3
ICI	11	15	26	21	17	38	164	23.2			
Sandoz	16	7	23	41	16	57	473	12.1			
Ciba-Geigy	9	13	22	9	17	26	125	20.8			
DowElanco											
Dow AgroSciences											
DuPont				0	20	20					
Pioneer	30	5	35	46	7	53	840	6.3	133	1,60 0	8.3
Monsanto	1	15	16	1	15	16					
DeKalb/Pfizer	19	6	25	16	6	22	205	10.7	41	387	10.6

Company	1986 Plant breeding			1988-89 Plant breeding					1996 Seed		
	Conventional breeding R&D	**Plant biotechnology R&D**	**Total plant breeding R&D**	**Conventional breeding R&D**	**Plant biotechnology R&D**	**Total plant breeding R&D**	**Seed sales (fiscal 1989)**	**R&D as share of seed sales**	**Seed R&D**	**Annual revenue**	**R&D as share of revenue**
		Million current dollars			*Million current dollars*			*Percent*	*Million current dollars*		*Percent*
Cargill[1]							250		37	250	14.8
Limagrain				22	5	27	335	8.1	60	660	9.1
KWS				18	5	23	151	15.2	47	350	13.4
Enimont				0	15	15					
Seminis									47	380	12.4
Shell	9	3	12	19	3	22					
Upjohn				24	3	27	270	10.0			

[1]Cargill'sinternational seedbusiness was purchasedby Monsanto, whileits U.S.seedbusiness was purchasedby AgrEvo.

Sources:1986:Grossmanet al.(1988, p.13), 1988-89:Fox (1990, p.42-43), 1996:James (1997, p.6).

Table 28. Research expenditures on crop improvement for corn, public and private

Expenditure type sector/type	1960	1965	1970	1975	1979	1984
	Million 1984 dollars					
Research expenditures -breeding:						
Private sector	11.0	13.4	18.8	28.9	43.6	59.2
Public sector	16.5	18.5	21.2	26.0	27.5	36.8
Total public and private	27.5	31.9	40	54.9	71.1	96
	Percent					
Share of total:						
Private	40.0	42.0	47.0	52.6	61.3	61.7
Public	60.0	58.0	53.0	47.4	38.7	38.3

Source:Huffman and Evenson (1993, p.159).

Table 29. Research expenditures on crop improvement for soybeans, public and private

Expenditure type sector/type	1960	1965	1970	1975	1979	1984
	Million 1984 dollars					
Research expenditures -breeding:						
Private sector	0.01	0.2	1.0	5.9	9.5	13.2
Public sector	9.79	10.9	14.7	23.9	40.6	41.9
Total public and private	9.80	11.1	15.7	29.8	50.1	55.1
	Percent					
Share of total:						
Private	0.10	1.8	6.4	19.8	19.0	24.0
Public	99.90	98.2	93.6	80.2	81.0	76.0

Source: Huffman and Evenson (1993, p.165).

With the development of commercially viable hybrids in the 1930s, corn was the first crop subject to the rapid shift from public to private R&D sector expenditures in plant breeding. Private seed companies accounted for 40 percent of total R&D spent on corn in 1960 and more than 60 percent in 1984 (table 28). The shift of R&D expenditures from the public to the private sector occurred more recently with soybeans, and may be partly credited to the PVPA (Fuglie et al., 1996, p. 53). In 1960, less than 1 percent of R&D expenditures on soybean improvement came from the private sector (table 29). By 1984, this share had risen to almost a quarter of the total (public and private) R&D spent on soybeans. Among private R&D expenditures on major crops, the share of R&D spent on soybeans by the 14 largest seed firms grew the fastest, from 1 percent in 1970 to 11 percent in 1980 (table 30). Public varieties of wheat and, to a lesser degree, cotton have been important sources of new seed for farmers. In addition, the public sector has continued to be the primary source of R&D investment and variety development for many small grains, such as oats, barley, and other minor field crops (Fuglie et al., 1996, pp. 53-55; Heisey, 1999, p. 19).

Table 30. Private research and development expenditures of the 14 largest seed firms, by year and crop

Year	Expenditures	Corn	Soybeans	Alfalfa	Wheat	Cotton
	Million current dollars	*Percent*				
1970	3.40	79	1	9	7	4
1972	3.94	77	2	8	9	5
1974	5.59	75	6	7	8	4
1976	8.39	73	7	6	11	3
Year	**Expenditures**	**Corn**	**Soybeans**	**Alfalfa**	**Wheat**	**Cotton**
	Million current dollars	*Percent*				
1978	10.26	71	10	6	9	4
1980	12.13	71	11	7	9	3

Source: Bsoutler and Marion (1985), p.31

Research Patterns in Terms of Scientist Years

Measuring the number of scientist years (SY) and funding per scientist (holding Ph.D. or M.S. degrees) allocated to specific areas of research offers additional insights into the public and private R&D effort on plant breeding. Across all crop varieties, the number of SY engaged in private plant breeding of major crops increased by 114 percent (from 701 to 1,498) between 1982 and 1994, while funding per scientist year had an apparent increase of 38 percent from $164,000 to $226,000 in current dollars over the same period (table 31), a 28-percent decrease in real terms. The distribution of SY between different crops has closely followed the allocation of R&D expenditures in dollar terms for both the public and private sectors (Figure 15). For instance, the private sector provided 94 percent of the total 545 SY allocated to corn breeding research in 1994, a fact that reflects the private sector's dominance in corn research. For such crops as cotton and soybeans, for which the public sector still plays a role in plant breeding and germplasm research, the private sector provided 77 percent (cotton) and 65 percent (soybeans) of the total SY allocated to plant breeding research in 1994 (table 32). Private sector research on wheat is even more limited: 41 percent of the total SY allocated to wheat breeding research came from the private sector in 1994.

A cost comparison of public and private R&D on plant breeding per scientist year for 1994 shows that, on average, expenditures in plant breeding were higher in the public sector ($286,840 per SY) than in the private sector ($225,898 per SY) (table 33). However, public R&D expenditures were lower than expenditures of the larger private firms ($290,000 per SY), likely due to the more complex nature of plant breeding R&D employed more than twice as many SY as the public (including the use of biotechnology techniques, which sector (1,498 versus 743), and total private sector is expensive, Kalton el al., 1989) carried out by both expenditures in plant breeding ($338 million) exceeded public sector and large firms. Overall, the private sector employed more than twice as many SY as the public sector (1,498 versus 743), and total private sector expenditures in plant breeding ($338 million) exceeded those of the public sector ($213 million) in 1994.

INTRODUCTIONS AND TRIALS OF NEW VARIETIES ARE INCREASING OVER TIME

The emergence of the private sector marks a significant shift in agricultural R&D. Decisions about R&D made in a competitive environment are likely to differ from those made in a public sector setting, where most R&D traditionally has taken place. Therefore, the amount of R&D, the emphasis of research, and other aspects of R&D are likely to change. In addition, public research priorities may change in response to R&D efforts undertaken by private firms: Public institutions may pursue initiatives that complement private efforts as well as research areas that the private sector neglects. Measures of plant breeding research output further highlight the shift in emphasis from public to private R&D.

Table 31. Number of firms and scientist years (SY) engaged in private plant breeding for major field crops

Crop	1982				1989*				1994			
	Number of firms	Share of firms	Scientist years	Share of SY	Number of firms	Share of firms	Scientist years	Share of SY	Number of firms	Share of firms	Scientist years	Share of SY
		Percent		*Percent*		*Percent*		*Percent*		*Percent*		*Percent*
Corn	66	24.54	255	36.38	75	27.57	371	34.77	91	27.66	510	34.05
Soybeans	26	9.67	52	7.42	34	12.50	86	8.06	38	11.55	101	6.74
Cotton	13	4.83	28	3.99	11	4.04	17	1.59	35	10.64	103	6.88
Wheat	21	7.81	42	5.99	11	4.04	47	4.40	27	8.21	54	3.60
Others	143	53.16	324	46.22	141	51.84	546	51.17	138	41.95	730	48.73
Total	269		701		272		1,067		329		1,498	
Dollars ($Mil.)114.95					272				338.462			
Dollars ($)/S Y163,980					306,306				225,898			

* 1989 figures for million dollars per SY are based on the average of figures for 1988 and 1990.

Source: Companies and Expenditures for 1994:Frey (1996), p.19; SY by crop, Frey (1996), pp.36-38; Companies and expendituresfor 1982 and 1989: Kaltonet al.(1990), p.24.

Table 32. Number of scientist years (SY) devoted to plant breeding, public and private, by crop, 1994

Crop/Crop category	Public sector		Private sector		Total	
	Number of SY employed	Share of total for the crop	Number of SY employed	Share of total for the crop	Number of SY employed	Share of total SY
		Percent		*Percent*		*Percent*
Corn	35	6.48	510	93.52	545	24.72
Soybeans	55	35.01	101	64.99	156	7.07
Cotton	31	22.94	103	77.06	134	6.09
Wheat	76	58.63	54	41.37	130	5.91
Other cereal crops	77	35.48	139	64.06	217	9.84
Other grain legumes	26	50.98	25	49.02	51	2.31
Other fiber crops	2	100.00	0	0.00	2	0.09
Forage	71	58.20	51	41.80	122	5.53
Fruit vegetable	46	21.60	167	78.40	213	9.66
Other crops	287	45.27	348	54.89	634	28.75
Total	706		1,499		2,205	

Source: Frey (1996), pp.6-11.

Table 33. Public and private research in plant breeding, scientist years (SY), and cost, 1994

Sector/Institution	Number of companies	Total number of SY	Cost per SY	Dollar input per sector/institution
			Dollars	*Million current dollars*
Private	329	1,498	225,898	338.5
Public	NA	743	286,840	213.2
ARS/USDA	NA	177	300,000	53.1
SAES	NA	530	293,500	155.5
Plant materials center	NA	36	125,000	4.5
Total	NA	2,241		551.6

* Average cost. Cost varies with company size, ranging from $148,000 for the smallest firm size category to $290,000 for the larger firms (Frey, 1996, p.19)NA = not available. SAES = State Agricultural Experiment Station.

Source: Frey (1996, p.19).

PVP Certificates

Plant variety protection certificates approved by USDA's PVPO are similar to patents issued for crop varieties. PVP certificates are intended to benefit both consumers and producers of improved crop varieties. Individual and corporate consumers benefit from the improved quality of agricultural goods that they use directly for consumption purposes or as inputs to the production of goods, such as livestock and medicine. And to the extent that new varieties increase productivity and supply, consumers also stand to gain from decreases in price. Seed producers, on the other hand, benefit from the exclusive rights they secure over the purity, breeding, marketing, distribution, and sales of improved varieties, allowing them to obtain a return on their investment of research and development resources (USDA, AMS, 2000).

Estimates of the time involved in producing new varieties in a breeding program range from 10 to 15 years to produce a marketable product (tables 34-35).[13] On an annual basis, a small breeding program was estimated to cost approximately $250,000 in the late 1980s, a sum adequate to cover the costs of a chief breeder, a staff of three or four, equipment, facilities and land (McMullen, 1987, p. 58). Even where larger firms realize economies of scale and scope in producing multiple varieties, the estimated development costs of a new variety range between $2.0 million and $2.5 million for the same period (McMullen, 1987, pp. 58-60). Given the magnitude of these investments, it is unlikely that plant breeders would have made this type of R&D investment without property rights protection.

The number of PVP certificates issued by the PVPO provides a useful indicator of the results of plant breeding research efforts. Research findings differ in showing the PVPA's effect on creating private sector incentives in research. Butler and Marion (1985, p. 79) conclude that during the 1970s, the PVPA resulted in "modest private and public benefits at modest private and public costs." Perrin et al. (1983), on the other hand, conclude that the act led to increased private investment in plant breeding for soybeans and other nonhybrid seed crops. These competing conclusions highlight the significance of time lags between R&D investment and the release of a new variety. For many varieties, this lag is often more than 10 years (table 35), implying that the economic returns to new R&D investments in plant breeding made in the 1970s were not fully captured in the studies by Butler and Marion and Perrin et al. (Fuglie et al., 1996, p. 38).

Table 34. Stages and time required in plant breeding

Stage	Hybrid	Open-pollinated
	Years	
Recognition	0	0
Parent-line preparation	4-5	0-2
Initial crosses	5-6	0-3
Progeny selection	6-10	3-11
Crop evaluation	7-12	5-12
Testing the variety	7-15	6-15
Determination of a new variety	8-12	7-13
Market evaluation	8-13	8-14
Application for plan variety protetection	8-14	9-14
Multiplication from individual plants or ears	9-14	9-15
Certification	9-14	9-15
Market introduction	10-15	10-17
Market acceptance	12-18	12-19
Market growth	13-19	13-20
Obsolescence	20-25	20-25

Source: McMullen (1987b), p.58.

Table 35. Research time required for developing new seed varieties, major field crops

Crop	Time required			Number	of varieties[1]
	From date of cross to date of determination	From date of determination to application date	Total	Cross to date Of determination	Date of determination to application
	Years				
Corn	5.5	2.0	7.5	4	6
Soybean	6.2	3.0	9.2	64	75
Cotton	8.0	4.2	12.2	27	57
Wheat	8.0	2.8	10.8	36	56
Rice	6.0	2.8	8.8	5	12
Average/total[2]	7.9	3.2	11.1	253	391

[1]Applicants are required to list the date of variety determination and date of application when submitting protection applications but they are not required to list the date the cross was made. For this reason, there are fewer varieties listed in the "Cross to date of determination" column.

[2]Averageand total include other field crops and vegetables not shown here.

Source: McMullen (1987), p.60, using data from Asgrow Seed Company.

The number of PVP certificates issued by the PVPO has grown rapidly since the 1970 PVPA after accounting for the time lag in plant breeding R&D (tables 36-37). This growth indicates the PVPA's positive effect on generating private sector incentives for plant breeding R&D. The increases were most marked for soybeans and corn, which together account for more than half of all certificates issued for field crops. By the end of 2002, 2,584 certificates were issued for varieties of U.S. origin for the four major field crops, including 1,078 for soybeans, 648 for corn, 568 for wheat, and 290 for cotton (table 37).

The majority of PVP certificates—about 84 percent—are held by the private sector. Among PVPs for major crops, the private sector owns close to 100 percent of corn certificates, 87 percent of cotton certificates, 84 percent of soybean certificates, and 68 percent of wheat certificates. Figure 16 captures the growth in PVP certificates issued for U.S. private and public entities for corn, soybeans, cotton, and wheat between 1970 and 2002.

Agricultural Biotechnology R&D

A driving force behind some of the increase in private sector R&D has been the introduction of modern biotechnology to agriculture. Biotechnology, broadly defined, is the application of biological science to affect living things. Under this broad definition, the entire endeavor of agricultural experimentation over thousands of years of human history might be considered "biotechnology." But the 20th century discovery of DNA and subsequent scientific advances have ushered in a new period of biological research. The application of genetic science to plants and animals in light of these discoveries is the sense in which this report employs the word "biotechnology."

The emergence of modern biotechnology is consistent with the more recent focus on plant breeding in private sector agricultural R&D (tables 25-

26; figures 13-14). In addition to techniques of modern biotechnology, the creation of new plant varieties with useful agronomic properties requires significant knowledge of plant breeding. In this sense, plant breeding and biotechnology are complementary. Moreover, the commercial success of GE crop varieties typically requires that biotechnology-derived trait enhancements are incorporated into successful cultivars. Acquisition of firms with established varieties by companies with the ability to improve varieties using biotechnology is one possible rationale for recent consolidation in the U.S. seed industry.

The rapid commercial success of GE varieties provides a preliminary measure of the technical success of the R&D efforts.[14] The number of field releases of plant varieties for testing purposes provides an ex-ante measure of R&D output.[15]

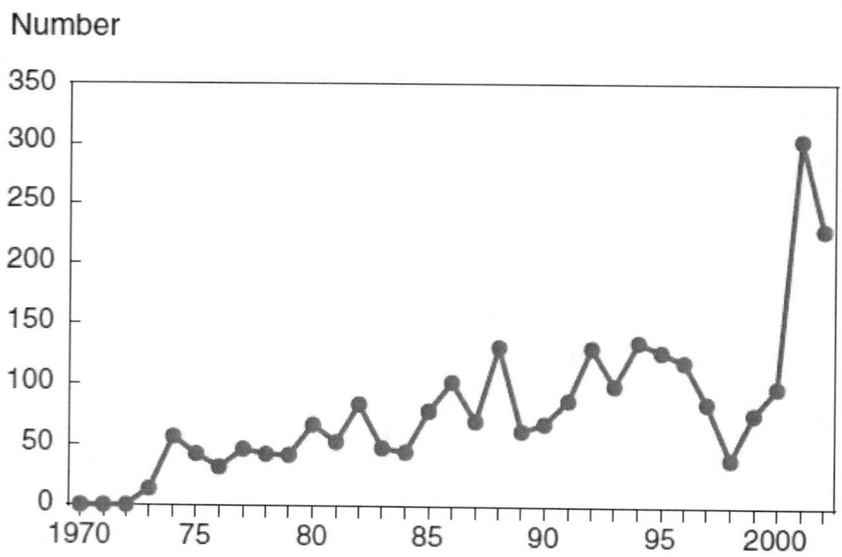

Source: Data source provided in table 37.

Figure 16. Number of PVP certificates issued for corn, soybeans, cotton, and wheat Number.

Table 36. Plant variety protection certificates issued for new crop varieties, field crops

Crop	Number of certificates issued							Share of certificate ownership	
	1971-74	1975-78	1979-82	1983-86	1987-90	1991-94	Total	Private	Public
								Percent	
Soybeans	34	69	132	150	114	162	661	0.84	0.16
Corn	0	1	6	50	104	161	322	1.00	0.00
Wheat	12	52	59	30	74	87	314	0.68	0.32
Cotton[1]	24	35	41	38	34	39	211	0.87	0.13
Subtotal for major crops	70	157	238	268	326	449	1,508	0.85	0.15
Barley	0	12	2	22	6	35	77	0.82	0.18
Beans, field	0	1	5	18	10	28	62	0.77	0.23
Oats	0	10	6	0	9	8	33	0.36	0.64
Rice	0	8	4	2	5	15	34	1.00	0.00
Sorghum	0	0	0	0	2	31	33	1.00	0.00
Canola	0	0	0	2	8	15	25	0.72	0.28
Safflower	0	3	2	1	5	6	17	0.88	0.12
Other field crops	0	16	15	13	18	13	75	0.85	0.15
Subtotal for other crops	0	50	34	58	63	151	356	0.80	0.20
Total field crops	70	207	272	326	389	600	1,864	0.84	0.16

[1] Figures for PVP certificates issued for cotton varieties given here may vary from figures presented else where in this report due to PVPO revisions and updates to PVP certificate listings.

Source: Fuglie (1996), p.38.

Field Releases

The process by which new GE varieties of organisms are released into the environment is regulated and monitored by USDA's Animal and Plant Health Inspection Service (APHIS). Private companies and public institutions proposing tests of such organisms in the environment either notify APHIS of their intent, in accordance with APHIS's field release notification procedures, or submit an application for a field release permit (referred to here as an application). If an APHIS review of the application (notification or permit application) establishes that there are no significant environmental risks associated with a release, a notification is acknowledged or a field permit is issued (referred to here as an "approval"), thereby allowing the breeder to pursue testing. Between 1987 and June 2001, APHIS received over 7,600 applications for field releases of GE varieties. Of these applications, APHIS approved the field release of more than 6,700 new varieties (table 38). Also significant is the annual growth of field release applications during this period: applications received annually by APHIS increased from just 9 in 1987 to a high of 1,206 in 1998 (Figure 17). Although some applications were denied or withdrawn, a significant majority—almost 90 percent—were approved by APHIS (table 39, Figure 18).

The majority of applications for field releases received from private companies and public institutions are for testing improved varieties of major crops. By mid-2001, applications received included more than 3,327 for corn varieties, 761 for potatoes, 601 for soybeans, 532 for tomatoes, 481 for cotton, and 209 for wheat (table 40). APHIS classifies each variety according to the variety's unique genetic characteristic, or phenotype, which distinguishes it from other varieties. Field release applications received by APHIS between 1987 and 2000 included varieties with such characteristics as herbicide tolerance (27 percent), insect resistance (25 percent), product quality usually associated with added value output traits (17 percent), virus resistance (9 percent), and agronomic properties (6 percent) (table 41). A small share of the applications for field releases also contain multiple or "stacked" traits, such as herbicide tolerance and insect resistance.

Table 37. PVP certificates issued for major field crops[1]

Calendar year issued	Corn					Soybeans					Cotton	Wheat	All major field crops
	Private origin	Public origin	Total U.S.	Foreign origin	Total	Private origin	Public origin	Total U.S.	Foreign origin	Total	Total U.S.	Total U.S.	Total
	Number												
1970	0	0	0	0	0	0	0	0	0	0	0	0	0
1971	0	0	0	0	0	0	0	0	0	0	0	0	0
1972	0	0	0	0	0	0	0	0	0	0	0	0	0
1973	0	0	0	0	0	10	3	13	0	13	0	1	14
1974	0	0	0	0	0	21	1	22	0	22	24	11	57
1975	0	0	0	0	0	8	5	13	0	13	14	16	43
1976	0	0	0	0	0	14	0	14	0	14	7	11	32
1977	3	0	3	0	3	21	1	22	0	22	5	17	47
1978	0	0	0	0	0	20	2	22	0	22	9	12	43
1979	0	0	0	0	0	19	2	21	0	21	11	10	42
1980	6	0	6	0	6	30	13	43	0	43	12	6	67
1981	1	0	1	0	1	30	2	32	0	32	10	10	53
1982	2	0	2	0	2	32	7	39	0	39	10	33	84
1983	3	0	3	0	3	33	1	34	0	34	11	0	48
1984	11	0	11	0	11	23	4	27	0	27	7	0	45
1985	8	0	8	0	8	37	11	48	0	48	8	14	78
1986	29	0	29	0	29	42	2	44	0	44	13	16	102

Table 37. (Continued)

Calendar year issued	Corn					Soybeans					Cotton	Wheat	All major field crops
	Private origin	Public origin	Total U.S.	Foreign origin	Total	Private origin	Public origin	Total U.S.	Foreign origin	Total	Total U.S.	Total U.S.	Total
						Number							
1987	15	0	15	1	16	36	5	41	0	41	11	2	70
1988	33	0	33	0	33	34	14	48	0	48	8	42	131
1989	19	0	19	0	19	12	3	15	0	15	0	28	62
1990	37	0	37	0	37	10	0	10	0	10	15	6	68
1991	35	0	35	0	35	33	4	37	0	37	14	1	87
1992	54	0	54	1	55	47	17	64	1	65	1	9	130
1993	36	0	36	5	41	6	0	6	0	6	4	48	99
1994	29	0	29	0	29	42	12	54	0	54	20	32	135
1995	22	0	22	0	22	58	15	73	0	73	6	26	127
1996	35	0	35	5	40	33	5	38	0	38	7	34	119
1997	30	0	30	10	40	24	4	28	0	28	4	13	85
1998	22	0	22	0	22	13	1	14	0	14	0	3	39
1999	26	0	26	0	26	16	4	20	0	20	0	30	76
2000	18	1	19	0	19	43	9	52	0	52	0	27	98
2001	88	4	92	3	95	124	10	134	0	134	10	64	303
2002	79	2	81	0	81	43	7	50	2	52	49	46	228
Total	641	7	648	25	673	914	164	1,078	3	1,081	290	568	2,612

[1] Figures for PVP certificates issued may vary from figures presented else where due to PVPO revision sand updates to PVP listings.

Source: Strachan (2003).

Table 38. Applications for field releases received by APHIS, by crop and year[1]

Crop	1987	1988	1989	1990	1991	1992	1993	1994	1995	1996	1997	1998	1999	2000	2001	Total (1987-2001)[1]
								Number								
Corn	0	0	0	3	15	44	134	262	348	279	406	563	385	420	468	3,327
Soybeans	0	0	4	5	6	36	68	69	62	52	55	85	68	51	40	601
Cotton	0	1	5	11	17	4	23	43	63	33	50	47	53	79	52	481
Wheat	0	0	0	0	0	0	0	3	4	12	16	20	39	65	50	209
Other	9	17	29	39	69	77	149	231	229	278	281	491	516	397	246	3,058
Total	9	18	38	58	107	161	374	608	706	654	808	1,206	1,061	1,012	856	7,676

[1] Includes field release notification sreceived by APHIS, and field release permitseitherissued, withdrawn, ordeniedby APHIS,between June 16, 1987,and July 9, 2001.

Source: Virginia Polytechnic Institute (2001).

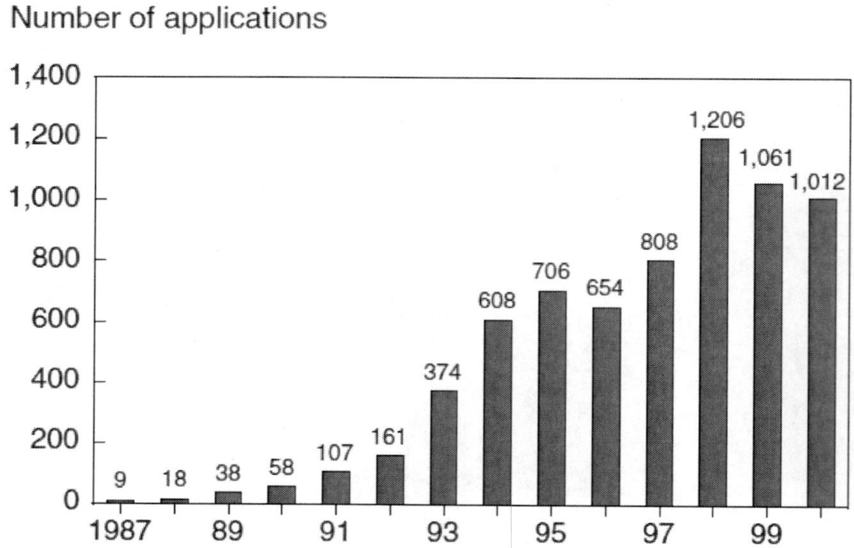

Source: Data source provided in table 38.

Figure 17. Total number of field release applications, by year.

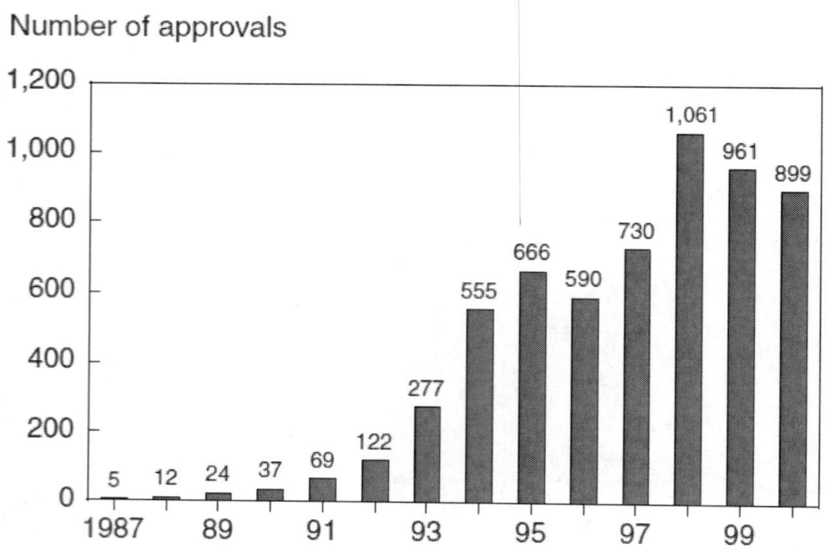

Source: Data source provided in table 39.

Figure 18. Total number of field release approvals, by year.

The breakdown of the number of applications for field releases by company is provided in table 42. Much more detail, by year and for each of the major crops —corn, soybeans, cotton, and wheat—is shown in tables 43-47.

Table 39. Status of field release notifications and permits[1]

Year	Received	Approved	Delayed approval	Denied	Withdrawn	Void	Pending	Total
				Number				
1987	9	5	4	0	0	0	0	18
1988	18	12	6	0	0	0	0	36
1989	38	24	14	0	0	0	0	76
1990	58	37	21	0	0	0	0	116
1991	107	69	38	0	0	0	0	214
1992	161	122	28	0	11	0	0	322
1993	374	277	29	0	68	0	0	748
1994	608	555	38	6	9	0	0	1,216
1995	706	666	14	2	18	5	0	1,411
1996	654	590	35	8	20	0	0	1,307
1997	808	730	13	33	28	3	0	1,615
1998	1,206	1,061	25	108	10	2	0	2,412
1999	1,061	961	25	46	22	6	0	2,121
2000	1,012	899	36	57	16	1	2	2,023
2001	856	777	0	25	13	1	39	1,711
Total[2]	7,676	6,785	326	285	215	18	41	15,346

[1] From June 16, 1987, to July 9, 2001.

[2] The total number of field releases approved by APHIS includes 961 permits issued and 6,156 notifications acknowledged.

Acknowledgments are given under the APHIS notification procedure, while issued refers to release permits issued by APHIS.

For most purposes, there is no difference in these two categories, and together, they equal the number of field releases approved by APHIS.

Source: Virginia Polytechnic Institute (2001).

Table 40. Number of applications for field releases, by crop[1]

Crop	2000		1987-2001	
	Number	*Percent*	*Number*	*Percent*
Corn	420	41.5	3,327	43.3
Potatoes	70	6.9	761	9.9
Soybeans	51	5.0	601	7.8
Tomatoes	25	2.5	532	6.9
Cotton	79	7.8	481	6.3
Wheat	65	6.4	209	2.7
Tobacco	15	1.5	202	2.6
Rapeseed	16	1.6	177	2.3
Rice	19	1.9	134	1.7
Beet	22	2.2	121	1.6
Melons	13	1.3	140	1.8
Other	217	21.4	991	12.9
Total	1,012	100.0	7,676	100.0

[1] From June 16, 1987, to July 9, 2001. Includes field release notifications received by APHIS, and field release permits either issued, with-drawn, or denied by APHIS between June 16, 1987, and July 9, 2001.

Source: Virginia Polytechnic Institute (2001).

Table 41. Share of applications for field releases received by APHIS, by trait

Trait	2000	1987-2001[1]
	Percent	
Agronomic properties	6.5	6.0
Bacterial resistance	1.3	1.2
Fungal resistance	5.5	5.5
Herbicide tolerance	30.1	27.4
Insect resistance	26.0	25.3
Marker gene	5.4	3.8
Nematode resistance	0.2	0.2
Product quality	11.5	17.3
Virus resistance	6.1	9.1
Other	7.3	4.4
Total[1]	100.0	100.0

[1] From June 16, 1987, to June 25, 2001. Note that certain products contain multiple or "stacked" traits and are thus included separately under each appropriate trait.

Source: Virginia Polytechnic Institute (2001).

Table 42. Status of notifications and field release permits by company, 1987-2001[1]

Company/ institution	Acknowledged[2]	Issued[2]	Pending	Denied	Withdrawn	Void	Total
		Number					
Monsanto	2,142	155	11	68	47	6	2,429
Pioneer	535	55	2	31	22	1	646
AgrEvo	312	14		14	4		344
DuPont	305	15			1		321
ARS	130	42	1	12	8	3	196
DeKalb	172	9		8	3		192
Calgene	90	74		2	8	1	175
Semnis Vegetable Seed	144	18	2	3	1		168
DNA Plant Tech	74	15			2		91
Northrup-King	69	11		3	5		88
University of Idaho	66	14		6	1	1	88
Upjohn	10	63			12		85
Aventis	72	4	1		8		85
Iowa State University	69	5		7	2		83
Asgrow	49	26		1	5		81
Novartis Seeds	74	3		2			79
ProdiGene	50	22	2	1	1		76
Stine Biotechnology	71			4			75
Rutgers University	59	12		4			75
Cargill	54	11		1	5		71
Dow	56	2	4				62
Agracetus	57	3		1			61
Agritope	47	6		6	1		60
Zeneca	47	2	1	7	1		58
Frito Lay	36	18		2	2		58
Other	1,366	362	17	102	76	6	1,929
Total	6,156	961	41	285	215	18	7,676

[1]From June 16, 1987, to July 9, 2001.

[2]Acknowledgments are given under the APHIS notification procedure, while issued refers to release permits issued by APHIS.

For most purposes, there is no difference in these two categories, and together, they equal the number of field releases approved by APHIS.

Source: Virginia Polytechnic Institute (2001).

Table 43. Field release approvals for corn, by company[1]

Crop	1987	1988	1989	1990	1991	1992	1993	1994	1995	1996	1997	1998	1999	2000	Total
							Number								
Monsanto					1	7	47	72	63	32	105	94	147	248	816
Pioneer					4	11	20	45	57	78	106	114	54	5	494
DuPont							12	67	63	14	17	19	7		199
AgrEvo							2	2	12		5	131	26		178
DeKalb				2	1	6	6	27	29	41	26	33	1		172
Northrup-King						3	6	10	42	10					71
ProdiGene											2	6	24	24	56
Novartis Seeds											12	22	7	12	53
Iowa State University									1	1	4	15	16	14	51
Stine Biotechnology											3	9	21	14	47
Plant Genetic Systems										4	42				46
Cargill					1	2	3	2	10	6	7	8	2		41
Holdens					2	4	3	5	14	7	2				37
Ciba-Geigy					2	3	4	5	10	12					36
Stanford University											7	4	12	13	36
Garst/ICI					1	2	2	4	3	5	5	8	3		33
Mycogen							1	2	1	1	6	8	11		30
Dow							1	1	2		1		6	18	29
Limagrain								1		10	6	6	1	4	28

Crop	1987	1988	1989	1990	1991	1992	1993	1994	1995	1996	1997	1998	1999	2000	Total
							Number								
Asgrow							1	4	6	3	3				17
Agracetus								1	5	5	3		1		15
NC+Hybrids									3	6	3	3			15
University of Minnesota									1	2	3	4	5		15
University of Arizona										1	1	5	3	4	14
Aventis														13	13
Golden Harvest Seeds									2	5	1	3	2		13
Rogers/Rogers NK							1	4	2	6					13
Rhone-Poulenc											1	3	6		10
Wyffels Hybrids										6	4				10
Southern Illinois Univ.										2	2	1	3	1	9
Upjohn					2	3	2								7
Zeneca											1	2	3	1	7
Other				1	1	2	3	6	11	16	5	9	5	15	74
Total	0	0	0	3	15	43	114	258	337	273	383	507	366	386	2,685

[1]From June 16, 1987, to December 31, 2000. Field release approvals are either categorized as notifications acknowledged by APHIS under its notification procedure, or field release permits issued by APHIS. For most purposes, there is no difference in the two categories, and together, they equal the number of field releases approved by APHIS.

Source: Virginia Polytechnic Institute (2001).

Table 44. Field release approvals for soybeans, by company[1]

Crop	1987	1988	1989	1990	1991	1992	1993	1994	1995	1996	1997	1998	1999	2000	Total
							Number								
Monsanto			4	4	3	22	13	12		1	6	20	38	21	144
DuPont							3	15	28	4	19	18	2	2	91
AgrEvo								6	16	16	8	13			59
Pioneer					1	4	11	16	5	3	7	7	2	2	58
Agracetus							1	4	9	13	2				29
Asgrow							1	6	2	9	5	6			29
Upjohn				1	2	5	7								15
Rhone-Poulenc											2	2	10		14
Aventis														10	10
DeKalb							2	1		2	1	3			9
Stine Biotechnology						1							1	6	8
Dairyland Seeds							2	3	1			1			7
Limagrain										1	1	3	1		6
University of Kentucky									1				2	2	5
AgriPro						1	2	1							4
Northrup-King						1	2	1							4
University of Illinois										2		2			4
Calgene												3			3
Iowa State University												1	1	1	3

Crop	1987	1988	1989	1990	1991	1992	1993	1994	1995	1996	1997	1998	1999	2000	Total
							Number								
Jacob Hartz							1	2							3
University of Georgia											1	1	1		3
University of Nebraska												1		2	3
FFR Cooperative							1	1							2
Ohio State University													2		2
Other							4					1			5
Total	0	0	4	5	6	34	50	68	62	51	52	82	60	46	520

[1]From June 16, 1987, to December 31, 2000. Field release approvals are either categorized as notifications acknowledged by APHIS under its notification procedure, or field release permits issued by APHIS. For most purposes, there is no difference in the two categories, and together, they equal the number of field releases approved by APHIS.

Source: Virginia Polytechnic Institute (2001).

Table 45. Field release approvals for cotton, by company[1]

Crop	1987	1988	1989	1990	1991	1992	1993	1994	1995	1996	1997	1998	1999	2000	Total
							Number								
Monsanto			2	6	6		9	11	26	12	14	26	39	54	205
Calgene			2	4	9	4	4	4	4	5	11	1			48
Delta & Pine Land							2	11	13	6	4				36
AgrEvo											12	10	4		26
Dupont				1	2		2	8	7						20
Agracetus		1						1	5	7	1				15
Mycogen									2	1	2	4	3		12
Texas Tech University												5	3	4	12
Aventis														9	9
Dow														8	8
All-Tex								2							2
ARS										1			1		2
Boswell										1	1				2
Jacob Hartz									2						2
Novartis Seeds														2	2
United Agri Products								1	1						2
American Cyanamid							1								1

Crop	1987	1988	1989	1990	1991	1992	1993	1994	1995	1996	1997	1998	1999	2000	Total
Bowdoin College														1	1
							Number								
Brownfield								1							1
Chembred								1							1
Dunn								1							1
Miles							1								1
Northrup-King			1												1
SeedCo								1							1
Williams Seed								1							1
Other															0
Total	0	1	5	11	17	4	19	43	60	33	45	46	50	78	412

[1] From June 16, 1987, to December 31, 2000. Field release approvals are either categorized as notifications acknowledged by APHIS under its notification procedure, or field release permits issued by APHIS. For most purposes, there is no difference in the two categories, and together, they equal the number of field releases approved by APHIS.

Source: Virginia Polytechnic Institute (2001).

Table 46. Field release approvals for wheat, by company[1]

Crop	1987	1988	1989	1990	1991	1992	1993	1994	1995	1996	1997	1998	1999	2000	Total
							Number								
Monsanto								2	3	7	6	13	27	52	110
University of Idaho										1	3	3	3	3	13
Montana State Univ.												1	4	3	8
Novartis Seeds											2	1	3	1	7
ARS									1	2		1		1	5
Applied Phytologic s														2	2
AgrEvo								1							1
Syngenta														1	1
Cargill														1	1
Other															0
Total	0	0	0	0	0	0	0	3	4	10	11	19	37	64	148

[1] From June 16, 1987, to December 31, 2000. Field release approvals are either categorized as notifications acknowledged by APHIS under its notification procedure, or field release permits issued by APHIS. For most purposes, there is no difference in the two categories, and together, they equal the number of field releases approved by APHIS.

Source: Virginia Polytechnic Institute (2001).

Table 47. Field release approvals for corn, soybeans, cotton, and wheat, by company, 1987-2000[1]

Corn		Soybeans		Cotton		Wheat	
Company	No.	Company	No.	Company	No.	Company	No.
Monsanto	816	Monsanto	144	Monsanto	205	Monsanto	110
Pioneer	494	DuPont	91	Calgene	48	University of Idaho	13
DuPont	199	AgrEvo	59	Delta & Pine Land	36	Montana State Univ.	8
AgrEvo	178	Pioneer	58	AgrEvo	26	Novartis Seeds	7
DeKalb	172	Agracetus	29	DuPont	20	ARS	5
Northrup-King	71	Asgrow	29	Agracetus	15	Applied Phytologics	2
ProdiGene	56	Upjohn	15	Mycogen	12	AgrEvo	1
Novartis Seeds	53	Rhone-Poulenc	14	Texas Tech University	12	Syngenta	1
Iowa State University	51	Aventis	10	Aventis	9	Cargill	1
Stine Biotechnology	47	DeKalb	9	Dow	8	Other	0
Plant Genetic Systems	46	Stine Biotechnology	8	All-Tex	2		
Cargill	41	Dairyland Seeds	7	ARS	2		
Holdens	37	Limagrain	6	Boswell	2		
Ciba-Geigy	36	University of Kentucky	5	Jacob Hartz	2		
Stanford University	36	AgriPro	4	Novartis Seeds	2		
Garst/ICI	33	Northrup-King	4	United Agri Products	2		
Mycogen	30	University of Illinois	4	American Cyanamid	1		
Dow	29	Calgene	3	Bowdoin College	1		
Limagrain	28	Iowa State University	3	Brownfield	1		
Asgrow	17	Jacob Hartz	3	Chembred	1		
Agracetus	15	University of Georgia	3	Dunn	1		

Table 47. (Continued)

Corn		Soybeans		Cotton		Wheat	
Company	No.	Company	No.	Company	No.	Company	No.
NC+Hybrids	15	University of Nebraska	3	Miles	1		
University of Minnesota	15	FFR Cooperative	2	Northrup-King	1		
University of Arizona	14	Ohio State University	2	SeedCo	1		
Aventis	13	Other	5	Williams Seed	1		
Golden Harvest Seeds	13	Delta Pine & Land	1	Other	0		
Rogers/Rogers NK	13	Michigan State University	1				
Rhone-Poulenc	10	Midwest Oilseeds	1				
WyFFels Hybrids	10	Land O'Lakes	1				
Southern Illinois University	9	Other					
Upjohn	7						
Zeneca	7						
Other	74						
Total	2,685	Total	524	Total	412	Total	148

[1] From June 16, 1987, to December 31, 2000. Field release approvals are either categorized as notifications acknowledged by APHIS under its notification procedure, or field release permits issued by APHIS. For most purposes, there is no difference in the two categories, and together, they equal the number of field releases approved by APHIS.

Source: Virginia Polytechnic Institute (2001).

Determination of Nonregulated Status

In the United States, once new varieties are successfully field tested and the research is fully documented, breeders may apply for a "determination of nonregulated status" from APHIS. This determination, once approved, allows the variety to be produced and sold commercially (USDA, APHIS, 2000). Out of the thousands of plant varieties approved for field release, APHIS had received 79 petitions for deregulation status as of mid-2001 (table 48). Of these 79 petitions, APHIS granted nonregulated status to a total of 53. These new varieties no longer fall under Federal regulation and can be planted freely throughout the United States.

Of the varieties granted nonregulated status by APHIS, 18 are corn varieties, 12 are tomatoes, 5 are soybeans, and 5 are cotton. Thirty-six of percent of these nonregulated varieties have herbicide-tolerance traits, 20 percent have insect-resistance traits, and 19 percent have traits to improve product quality (table 49).

Adoption of Crop Biotechnology Products

Successful transfer of a targeted trait to an elite strain may take plant breeders many crop generations. Superior hybrid corn varieties, for example, were introduced in the early 1930s after more than 25 years of research. Once development is complete, adoption of these crops takes time as well. The share of corn acreage planted with hybrid corn in the U.S. grew from about 1 percent of total corn planted in 1933 to more than 95 percent by 1960 (Figure 2). The speed of adoption might depend on the success of marketing efforts, the ability of growers to adapt farming practices to take advantage of the new varieties, and the superiority of the new varieties to existing varieties. The speed of adoption of corn hybrids differed by region because plant breeders had to produce varieties compatible with local growing conditions (Griliches, 1957).

By comparison, both the development and the adoption of genetically engineered field crop varieties were even more rapid than the adoption of hybrid corn varieties. The relative speed with which new varieties can be developed using modern biotechnology gives biotechnology an advantage over other techniques. Notwithstanding, first-generation biotechnology products were commercially available for farmers in the mid-1990s after about 15 years of research and development. Following their release in 1996, U.S. farmers rapidly adopted GE crops with herbicide-tolerant and insect-resistant traits (Figure 3).

Table 48. Crops no longer regulated by USDA, 1987-2001[1]

Crop	Petitions submitted	Petitions approved	Share of petitions approved
	Number	*Number*	*Percent*
Corn	23	18	33.96
Tomato	23	12	22.64
Potato	23	5	9.43
Soybeans	23	5	9.43
Cotton	23	5	9.43
Other	23	8	15.09
Total	79	53	100.00

[1] From June 16, 1987, to June 25, 2001.
Source: Virginia Polytechnic Institute (2001).

Table 49. Crops no longer regulated by USDA, by trait, 1987-2001[1]

Trait	Petitioned	Approved	Share approved
	Number	*Number*	*Percent*
All crops:			
Agronomic properties (AP)	8	6	9.38
Herbicide tolerance (HT)	29	23	35.94
Insect resistance (IR)	22	13	20.31
Stacked (HT, IR)	5	4	6.25
Product quality (PQ)	18	12	18.75
Virus resistance (VR)	9	6	9.38
Total	91	64	100.00
Corn:			
Agronomic properties (AP)	3	3	16.67
Herbicide tolerance (HT)	8	7	38.89
Insect resistance (IR)	10	5	27.78
Stacked (HT, IR)	5	3	16.67
Total	26	18	100.00
Soybeans:			
Herbicide tolerance (HT)	5	4	80.00
Product quality (PQ)	2	1	20.00
Total	7	5	100.00
Cotton:			
Herbicide tolerance (HT)	3	3	60.00
Insect resistance (IR)	3	1	20.00
Stacked (HT, IR)	1	1	20.00
Total	7	5	100.00

[1] From June 16, 1987, to June 25, 2001.
Source: Virginia Polytechnic Institute (2001).

Table 50. Status of petitions for deregulation, by company, 1987-2001[1]

Company/institution	Approved	Incomplete	Pending	Withdrawn	Void	Total
			Number			
Monsanto	16		2	8		26
AgrEvo	10			3		13
Calgene	9				1	10
DeKalb	2			1		3
DuPont	2			1		3
DNA Plant Tech	1			1		2
Agritope	1			1		2
Bejo	1			1		2
Zeneca		1		1		2
Cornell University	1			1		2
Asgrow	1					1
Aventis				1		1
Ciba-Geigy	1					1
Dow	1					1
Upjohn	1					1
Syngenta				1		1
University of Saskatchewan	1					1
Zeneca & Petoseed	1					1
Mycogen				1		1
Northrup King	1					1
Novartis Seeds	1					1
Pioneer	1					1
Plant Genetic Systems	1					1
Vector Tobacco			1			1
Total	53	1	3	21	1	79

[1] From June 16, 1987, to June 25, 2001.

Source: Virginia Polytechnic Institute (2001).

Concentration and Private Sector R&D

A relatively small number of firms are active in the field of crop biotechnology, particularly with respect to major field crops (tables 43-46). Only a few firms have secured approval for unregulated release of genetically modified crops, a number likely affected by recent merger and acquisition activity.

As Fulton and Giannakas (2002) observe, the relevant measure of market concentration is not always based on output markets (sales). To assess the impact of mergers focusing on innovative activity, the Federal Trade Commission is using innovation competition.

To construct a measure of concentration in innovation activity for the case of crop biotechnology, this report uses the regulatory approvals of GE varieties. In particular, this section adapts the CR4 measure used to quantify concentration in seed sales. Figures 19-21 show the percentage of field releases obtained by the leading four firms during 1988-2000. The top four firms control well over 50 percent of these approvals, suggesting concentration in R&D as well as potential barriers to entry for potential competitors. Note that much of the concentration reflects mergers and acquisitions among firms listed individually in the tables.

Based on the four-firm concentration ratio of approvals, corn seed remains the least concentrated industry relative to the other three crops. This finding is fairly consistent with earlier findings presented on the four-firm concentration ratio of corn in terms of sales. Moreover, the level of corn seed R&D concentration has remained relatively constant, at around 72 percent since 1990, which is also consistent with earlier findings. Soybean seed R&D remains highly concentrated in terms of field release approvals, although this concentration generally follows a downward trend. Again, this is fairly consistent with trends for the four-firm concentration ratio when measured in more conventional terms earlier. Cottonseed R&D, on the other hand, shows a trend toward increasing concentration, from 89 percent in 1993 to 96 percent in 2000, a finding also consistent with earlier market concentration measures.[16] Finally, the wheat seed industry remains highly concentrated, although this may be due to the limited number of private firms operating in the sector; only two of the largest four institutions were private firms.

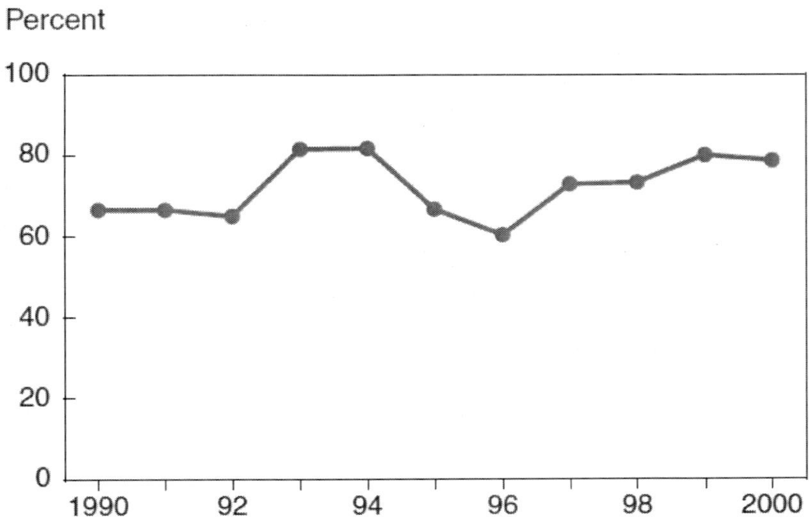

Source: Data source provided in table 43.

Figure 19. Four-firm concentration ratio in APHIS field release approvals for corn.

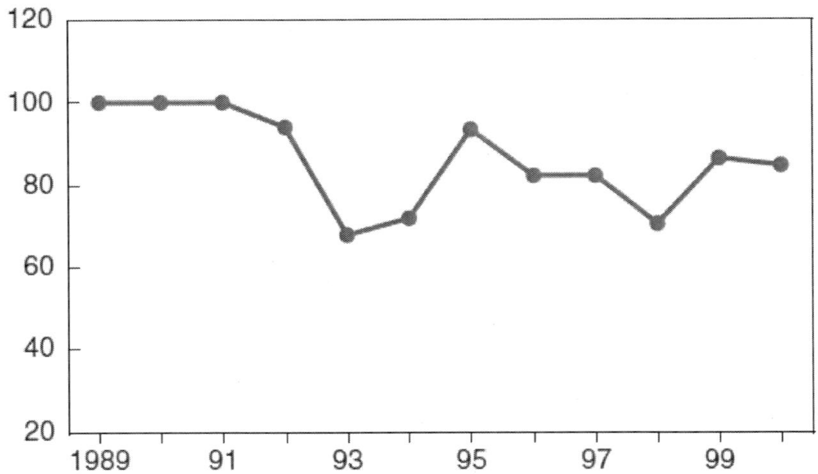

Source: Data source provided in table 44.

Figure 20. Four-firm concentration ratio in APHIS field release approvals for soybeans.

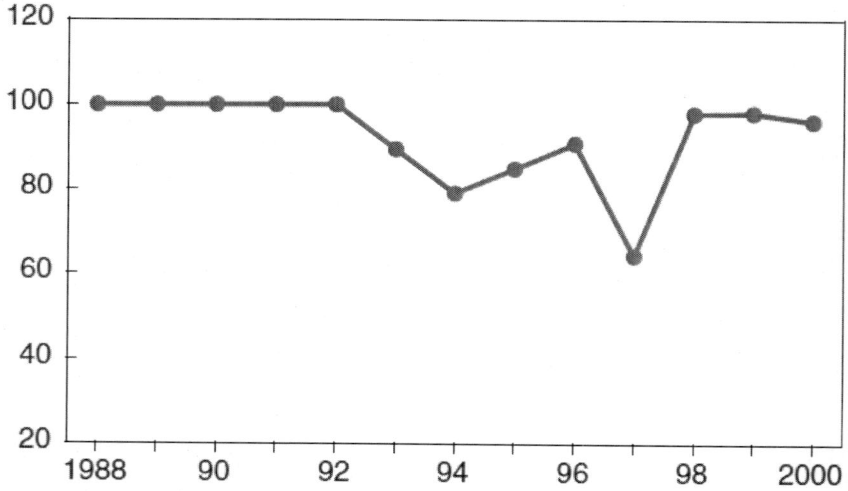

Source: Data source provided in table 45.

Figure 21. Four-firm concentration ratio in APHIS field release approvals for cotton.

Between 1988 and 2000, Monsanto received approvals for over 1,200 new varieties of corn, soybeans, cotton, and wheat, making it the leader in product development, followed by other large companies such as DuPont/ Pioneer, and AgrEvo. The majority of the permits issued to leading companies have been for testing of new corn varieties, followed by soybeans and cotton (table 47). Similarly, out of the 53 approvals for nonregulated status given by APHIS, 28 were for varieties developed by Monsanto/ Calgene/Asgrow/DeKalb, 10 were for AgrEvo, and 3 were for DuPont/Pioneer (table 50). Thus, the largest two firms hold 70 percent of the nonregulated varieties.

Patent ownership shows a pattern of concentration similar to that evident in other R&D measures. Most of the biotech patents awarded to private sector firms are held by a small number of large companies. As of 1996/97, DuPont/Pioneer held the largest number of patents for corn and soybeans, followed by Monsanto (table 51) (Brennan et al., 1999, p. 167).

As with regulatory approvals, the leading firms in the sector have received intellectual property rights not only through their R&D investment but also through recent mergers and acquisitions.

Table 51. Biotech patent ownership for corn and soybeans, by company

	Corn (1982-96)	Soybeans (up to 1997)	Corn and soybeans
	Number of patents held		
Aventis/Rhone-Poulenc-Agrochem	3	4	7
AgrEvo /PGS	4	3	7
Novartis/Syngenta	17	2	19
Zeneca	0	3	3
Dow Chemical/Mycogen	3	4	7
DuPont/Pioneer	28[1]	42[2]	70
Monsanto DeKalb	11[3]	23	34
Cyanamid	3	0	3
Others	69	49	118
Total	138	130	268

[1] Includes 21 from Pioneer.

[2] Includes 27 from Pioneer.

[3] Includes 4 from DeKalb.

Source: Derwent Biotechnology Abstracts, as reported in GRAIN(1996, 1997) and in Brennan et al.(2000).

REFERENCES

Abrahams, P. "ICI To Split and Shed 9,000 Jobs Worldwide: Investment in Petrochemicals Halted, Pounds 384m Pre-Tax Loss Reported." *Financial Times.*, February 26, 1993, p. 1.

Advanta, Facts and Figures. Accessed at URL: http://www.advantaseeds.com/show?id=11882& langid=96

AgBiotech Reporter. "Banner Year at D&PL." January 2000, p. 4. AgrEvo. Accessed at URL: http://www.archive.hoechst.com/english/life_sciences/ernaehrung/lw_zahlenfakten.html

Agrawal, PK; Agrawal, BD; Venkat, Rao, Jai, Singh. "Seed Multiplication, Conditioning and Storage." In Morris, Michael L; *Maize Seed Industries in Developing Countries.*, Boulder, CO: CIMMYT, 1998, 103-24.

Airy, John M; Tatum, LA; Sorenson, JW. Jr. "Producing Seed of Hybrid Corn and Grain Sorghum. In *Seeds: The Yearbook of Agriculture, 1961.* U.S. Department of Agriculture, 1961, 145-53.

Alston, JM; Pardey, PG. *Making Science Pay: The Economics of Agricultural R&D Policy*. Washington, DC: AEI Press, 1996.

Alston, JM; Venner, RJ. "The Effects of the U.S. Plant Variety Protection Act on Wheat Genetic Improvement." *Research Policy*, 31(2002), 527-42.

Appelbaum, E. "The Estimation of the Degree of Oligopoly Power." *Journal of Econometrics* 19(1982), 287-99.

Appelbaum, E. "Testing Price Taking Behavior." *Journal of Econometrics*, 9(1979), 283-94.

Asgrow Seed Company. "History." Accessed at URL: http://www.asgrow.com/ corporate/History.html, June 5, 2000.

Aventis Agriculture. "Introducing Aventis." Accessed at URL: http://www2.aventis.com/introduce/ agro.htm, June 5, 2000.

Azzam, A. "Measuring Market Power and Cost-Efficiency Effects of Industrial Concentration." *Journal of Industrial Economics*, 45(4)(1997), 377-86.

Azzam, AE. Pagoulatos. "Testing Oligopolistic and Oligopsonistic Behavior: An Application to the U.S. Meat Packing Industry." *Journal of Agricultural Economics*, 41(1990), 362-70.

Azzam, A; Schroeter. JR. "The Tradeoff Between Oligopsony Power and Cost Efficiency in Horizontal Consolidation: An Example From Beef Packing." *American Journal of Agricultural Economics*, 77(1995), 825-36.

Bayer CropScience. "Bayer CropScience Launch." Accessed at URL: http://www.bayercropscience.co. uk/home, September 27, 2002.

Beach, DE; Fernandez-Cornejo, J. "Setting Research Priorities in the Public Sector: A Suggested Framework for the AARC Center." *Journal of Agricultural Economics Research*, 45 (Summer 1994), 3-16.

Begemann, BD. "Competitive Strategies of Biotechnology Firms: Implications for U.S. Agriculture." *Journal of Agricultural and Applied Economics*, 29 (June 1997), 117-22.

Bjornson, B. "Capital Market Values of Agricultural Biotechnology: How High and Why?" *AgBioForum*, 1(2)(1998), 69-73. Accessed at URL: http://www.agbioforum.missouri.edu, September 22, 1999.

Blonigen, BA; Taylor. ChT. "R&D Intensity and Acquisitions in High-Technology Industries: Evidence From the U.S. Electronic and Electrical Equipment Industries." *Journal of Industrial Economics*, 48(1) (March 2000), 47-70.

Boswell, VR. "The Importance of Seeds." In *Seeds: The Yearbook of Agriculture, 1961*. U.S. Department of Agriculture, 1961, 1-10.

Brennan, MF; Pray, CE; Courtmanche. A. "Impact of Industry Concentration on Innovation in the U.S. Plant Biotech Industry." In Lesser, William H; ed; *Transitions in Agbiotech: Economics of Strategy and Policy. Proceedings of NE-165 Conference, June 24-25, 1999*, Washington, DC. Connecticut: University of Connecticut, 2000.

Butler, LJ; Marion, BW. *The Impacts of Patent Protection on the U.S. Seed Industry and Public Plant Breeding*. North Central Region Research Publication 304, North Central Project 117, Monograph 16. Madison, WI: Research Division, College of Agricultural and Life Sciences, University of Wisconsin, September 1985.

Carlton, DW; Perloff, JM. *Modern Industrial Organization*,2nd edition. Reading, MA: Addison-Wesley, 1994.

Chen, ZR. Lent. "Supply Analysis in an Oligopsony Model." *American Journal of Agricultural Economics*, 74(1992), 973-79.

Clark, A. "Zeneca Spends Pound 46m on Dutch Genetic Engineer." *Daily Telegraph.*, May 14, 1997, p. 29.

Cohen, WM; Levin, RC. "Empirical Studies of Innovation and Market Structure." In Schmalensee, R R.D. Willig, *Handbook of Industrial Organization*, Vol. II. Amsterdam: Elsevier, 1989.

Cosun. "Advanta Seed Technology." Accessed at URL: http://www. cosun.com/ advantagb.htm, June 5, 2000.

Cotton USA. Accessed at URL: http://www.cottonus.org/NewsResources/ SouthMid.htm; http://www.cottonus.org/NewsResources/West.htm, no date.

Cowling, KG; Waterson. M. "Price-Cost Margins and Market Structure." *Economica*, 43(1976), 267-74.

Daily Telegraph. "Zeneca's Seeds Business Merges." August 13, 1996, p. 25.

Day-Rubenstein, K; Heisey. P; "Agricultural Resources and Environmental Indicators: Crop Genetic Resources." Chapter 3.2 of *ERS Agricultural Resources and Environmental Indicators*, Agricultural Handbook No. AH722. April 2001. Accessed at URL: http://www.ers.usda.gov/publications/arei/ arei2001/arei3_2/AREI3_2geneticresources.pdf

Dow AgroSciences. "Company." Accessed at URL: http://www.dowagro.com/ main/company/default.htm, August 7, 2000.

Dow AgroSciences, News Release. "Dow Agro-Sciences To Reorganize, Eliminate 500 Positions Globally." September 19, 2002. Accessed at URL: http://www.dowagro.com/newsroom/news/091902.htm, September 27, 2002.

Duvick, DN. "Genetic Contributions to Yield Gains of U.S. Hybrid Maize, 1930 to 1980." In Fehr, W.R; ed; *Genetic Contributions to Yield Gains of Five Major Crop Plants*. Madison, WI: CSSA, 1984, 15-48.

Duvick, DN. "Genetic Rate of Gain in Hybrid Maize Yields During the Past 40 Years." *Maydica*, 22(1977),187-96.

Duvick, DN. Plant Breeding: Past Achievements and Expectations for the Future. *Economic Botany*, 40 (1992), 289-97.

Duvick, DN. "United States." In Morris, M.L; ed; *Maize Seed Industries in Developing Countries.*, Boulder, CO: CIMMYT, 1998, 193-211.

Emsweller, SL. "Fundamental Procedures in Breeding Crops." In *Seeds: The Yearbook of Agriculture, 1961*. U.S. *Department of Agriculture*, 1961, 127-33.

Falck-Zepeda, JG. Traxler. "The Role of Federal, State, and Private Institutions in Seed Technology Generation." Greg Traxler and José Falck-Zepeda. In Fuglie, K.O D.E. Schimmelpfennig, eds; *Public-Private Collaboration in Agricultural Research: New Institutional Arrangements and Economic Implications.*, Ames, IA: Iowa State University Press, 2000.

Fernandez-Cornejo, J. "Adoption of Genetically Engineered Crops in the U.S." Data product. U.S. Department of Agriculture, Economic Research Service, November 2003. Accessed at URL: http://www.ers.usda.gov/data/BiotechCrops

Fernandez-Cornejo, Shumway, JCR. "Research and Productivity in Mexican Agriculture." *American Journal of Agricultural Economics* 79(1997), 738-53.

Fernandez-Cornejo, Spielman, JD.*Concentration, Market Power, and Cost Efficiency in the Corn Seed Industry*. Selected paper presented at the 2002 AAEA meetings, Long Beach CA, July 28-31, 2002.

Fernandez-Cornejo, JWD. McBride. *Genetically Engineered Crops for Pest Management in U.S. Agriculture: Farm Level Effects*. Agricultural Economic Report 786. U.S. Department of Agriculture, Economic Research Service, April 2000.

Fernandez-Cornejo, JWD. McBride. *The Adoption of Bioengineered Crops*. Agricultural Economic Report 810. U.S. Department of Agriculture, Eco nomic *Research Service*, May 2002.

Fernandez-Cornejo, JWD. McBride. "Genetically Engineered Crops: U.S. Adoption and Impacts." *Agricultural Outlook*, No. 294, September 2002a, 24-27.

Fernandez-Cornejo, JD; Spielman, J. Keller. *Concentration, Market Power, and Cost Efficiency in the Cotton Seed Industry.* Selected paper presented at the 2002 SAEA meetings, *Orlando*, FL, Feb. 4-6, 2002.

Fernandez-Cornejo, J; Caswell, M; Klotz-Ingram. C. "Seeds of Change: From Hybrids to Genetically Modified Foods." *Choices* (4th quarter 1999), 18-22.

FIS/ASSINSEL (The International Seed Trade Federation / The International Association of Plant Breeders for the Protection of Plant Varieties). *Statistics*. Accessed at URL: *http://www.worldseed.org/statistics.html*, April 14, 2000.

Food and Agriculture Organization (FAO). "Agreement Reached on Protecting Plant Genetic Resources," July 16, 2001. Accessed at URL: http://www.fao.org/news/ 2001/010703-e.htm, July 30, 2001.

Fox, Patrick. *Biotechnology and the Seed Industry.* UK: PJB Publications, June 1990.

Frey Kenneth. J. *National Plant Breeding Study-1: Human and Financial Resources Devoted to Plant Breeding Research and Development in the United States in 1994*. Special Report 98. Ames, IA: Iowa Agriculture and Home Economics Experiment Station, *Iowa State University*, 1996.

Frisvold, G; Sullivan, J; Raneses, A. Who Gains From Genetic Improvements in U.S. Crops? *AgBio-Forum* 2 (3 & 4)(1999), 237-46. Accessed at URL: http://www.agbio.forum.org, July 17, 2000.

Fuglie, K; Ballenger, N; Day, K; Klotz, C; Ollinger, M; Reilly, J; Vasavada, U; Yee. J. *Agricultural Research and Development*. Acricultural Economic Report 735. U.S. Department of Agriculture, *Economic Research Service*, May 1996.

Fulton, MK. Giannakas. "Agricultural Biotechnology and Industry Structure." *AgBioForum*. 4(2),137-51. Accessed at URL: *http://www.agbioforum.com.*

Garst Seed Company. "Mike Muston Named Vice President for Advanta, Americas." Accessed at URL: http://www.garstseed.com/ articles/ pres990313-muston.htm, June 6, 2000.

Gollop, FM; Roberts. MJ. "Firm Interdependence in Oligopolistic Markets." *Journal of Econometrics* 10(1979), 313-31.

Gould, F. "Genetics of Plant Herbivore Systems: Interactions Between Applied and Basic Study." In Denno, R.F M.S. McClure, eds, *Variable Plants and Herbivores in Natural and Managed Systems*, New York: Academic Press, 1983.

Graff, GD; Rausser, GC; Small, AA. "Agricultural Biotechnology's Complementary Intellectual Assets." *Review of Economics and Statistics*, Forthcoming.

Grain. "The Biotech Battle Over the Golden Crop." *Seedling* 13(3) (October 1996). Accessed at URL: http://www.grain.org/publications/oct964-en-p.htm.

Grain. "Soybean: The Hidden Commodity." *Seedling* 14(2) (June 1997). Accessed at URL: http://www.grain.org/publications/oct973-en-p.htm.

Groosman, A; Linnemann, A; Wierma. H. *The International Dimension of Seed Industry Development: Draft Final Report of the Project*. Paper prepared for the seminar on "Technology Development and Changing Seed Supply Systems," Tilburg, 1988.

Hansen, L; Knudson, M. "Property Right Protection of Reproducible Genetic Material." *Review of Agricultural Economics* 18(1996), 403-14.

Hayenga, ML. "Structural Change in the Biotech Seed and Chemical Industrial Complex." *AgBioForum*, 1(2)(1998), 43-55. Accessed at URL: http://www.agbioforum.missouri.edu, September 22, 1999.

Heisey, Paul, W. *Public Sector Agricultural Research in an Era of Rapidly Expanding Private Sector Investment.* Paper prepared for the Western Economic Association International, Huntington Beach, CA, July 10, 1999.

Heisey, Paul, W. Private communication, September 23, 1999a.

Heisey, PW; Srinivasan, CS; Thirtle. C. *Public Sector Breeding in a Privatizing World*. Agricultural Economic Report 772. U.S. Department of Agricul ture, Economic Research Service, August 2001.

Hoechst Schering AgrEvo. "Agriculture." Accessed at URL: http://www.archive.hoechst.com/ english/life_sciences/ernaehrung/lw_agrevo. html and http://www. archive.hoechst.com/english/ life_sciences/ernaehrung /lw_kooperationen. html, August 7, 2000.

Huffman, WE; Evenson. RE. *Science for Agriculture*. Ames, IA: Iowa State University Press, 1993.

International Union for the Protection of New Varieties of Plants (UPOV). *UPOV Convention (1991 Act)*. Accessed at: URL: http://www. upov.int/eng/convntns/1991/w_up910_.htm#_1, July 24, 2001.

_________. "States Party to the International Convention and Addresses of Plant Variety Protection Offices." Accessed at URL: http://www.upov.int/eng/ ratif/index.htm, July 30, 2001a.

________. "The UPOV System of Plant Variety Protection." Accessed at URL: http://www.upov.int/ en/about/upov_system.htm, on July 9, 2003.

Iwata, G. "Measurement of Conjectural Variations in Oligopoly. *Econometrica* 42(1974), 947-66.

James, C. "Progressing Public-Private Sector Partnerships in International Agricultural Research and Development." *ISAAA Briefs*, 4(1997).

James, C; Krattiger. AF. "Global Review of the Field Testing and Commercialization of Transgenic Plants: 1986-1995: The First Decade of Crop Biotechnology." *ISAA Briefs* 1(1996).

Jenkins, MT. "Corn Improvement." In *Yearbook of Agriculture, 1962*. U.S. Department of Agriculture, 1936, 455-522.

Jewett, TO. "Thomas Jefferson: Father of Invention." *The Early America Review*, Vol. III, n. 1. (Winter 2000). Anna Maria, Florida. Downloaded from http://www.earlyamerica.com on December 17, 2001.

Joly, Pierre-Benoit, S. Lemarie "Industry Consolidation, Public Attitude and the Future of Plant Biotechnology in Europe." *AgBioForum* 1(2)(1998), 85-90. Accessed at URL: http://www.agbioforum. missouri.edu, September 22, 1999

Just, RE; Hueth. DL; "Multimarket Exploitation: The Case of Biotechnology and Chemicals." *American Journal of Agricultural Economics*, 75(4)(1993), 936-45.

Kalaitzandonakes, N. "Mycogen: Building a Seed Company for the Twenty-First Century." *Review of Agricultural Economics*, 19(2)(1997), 156-66.

Kalaitzandonakes, N; Bjornson, B. "Vertical and Horizontal Coordination in the Agrobiotechnology Industry: Evidence and Implications." *Journal of Agricultural and Applied Economics*, 29(4)(1997), 129-139.

Kalaitzandonakes, N; Hayenga, ML. *Structural Change in the Biotechnology and Seed Industrial Complex: Theory and Evidence.* Paper presented at the Transitions in AgBiotech: Economics of Strategy and Policy Conference, Washington, DC, June 24-25, 1999.

Kalton, RR; Richardson, PA; Frey. NM. "Inputs in Private Sector Plant Breeding and Biotechnology Research Programs in the USA." *Diversity* 5(1990), 22-25.

Kidd George, H. (L. William Teweles and Co.). *Ag Biotech* (statement). Proceedings of the Conference held in Arlington, VA, March 28-30, 1989.

Kimle, K; Hayenga. ML. "Structural Change Among Agricultural Input Industries." *Agribusiness* 9(1)(1993), 15-27.

Kimle, K; Hayenga. ML. *Agricultural Inputs and Processing Industries.* Ames, IA: Iowa State University, 1992.

King, John L. *Concentration and Technology in Agricultural Input Industries.* U.S. Department of Agriculture, Economic Research Service, Agriculture Information Bulletin 763, March 2001.

Klotz, Cassandra, Fuglie, K; and Pray, C. *Private-Sector Agricultural Research Expenditures in the United States, 1960-1992.* NRED-9525. U.S. Department of Agriculture, Economic Research Service, October 1995. Update for 1993-1997 released in July 1998.

Klotz-Ingram, C. Private communication. July 24, 2000.

Kloppenburg, JR. *First the Seed: A Social History of Plant Breeding and the Seed Industry in the United States.* Ph. D. dissertation, Cornell University, January 1985.

Knudson, MK. "The Role of the Public Sector in Applied Breeding R&D: The Case of Wheat in the U.S.A." *Food Policy* 15(3)(1990), 209-217.

Knudson, MK. Pray, CE. "Plant Variety Protection, Private Funding, and Public Sector Research Priorities." *American Journal of Agricultural Economics*, 73:3 (August 1991), 882-86.

Knudson, MK; Hansen, L. *Intellectual Property Rights and the Private Seed Industry.* Agricultural Economic Report 654. U.S. Department of Agriculture, Economic Research Service, November 1991.

Krull, DF; Prescott, JM; Crum, CW. "Seed Marketing and Distribution." In Morris, Michael, L; *Maize Seed Industries in Developing Countries.* Boulder, CO: CIMMYT, 1998, 125-41.

Leibenluft, RF. *Competition in Farm Inputs: An Examination of Four Industries.* Federal Trade Commission, Office of Policy Planning, February 1981.

Lerch, VD. "The Plant Variety Protection Act: A Current Look." *Seed World.* U.S. Department of Agriculture, Agricultural Marketing Service, Plant Variety Protection Office, 1989.

Lesser, W. "Intellectual Property Rights and Concentration in Agricultural Biotechnology." *AgBioForum* 1(2) (Fall 1998), 56-61. Accessed at URL: http:// www.agbioforum.missouri.edu, August 21, 2001.

Lucier, G; Chesley, A; Ahearn, M. *Farm Income Data: A Historical Perspective.* Statistical Bulletin 740. U.S. Department of Agriculture, Economic Research Service, 1986.

Maredia, MK; Byerlee, D. "Toward Efficiency Allocation of Research Resources in the Presence of Spillovers: Lessons From Wheat Iimprovement Research." In Maredia, M.K, and D.Byerlee, eds; *Global Wheat Improvement System: Prospects for Enhancing Efficiency in the Presence of Spillovers.* Mexico, DF (Mexico), CIMMYT. 1999. 118-26.

Maredia, MK; Ward. R. "Wheat Breeding Environments: A Conceptual and Empirical Analysis." In Maredia, M.K, and D. Byerlee, eds; *The Global Wheat Improvement System: Prospects for Enhancing Efficiency in the Presence of Spillovers*. Mexico, DF (Mexico), CIMMYT. 1999, 12-21.

McKay John, W. "How Seeds Are Formed." In *Seeds: The Yearbook of Agriculture, 1961*. U.S. Department of Agriculture, 1961, 11-17.

McMullen, N. *An Introduction to Seed Production.* NPA Report Series No 227. National Planning Association, 1987a.

McMullen, N. *Seeds and World Agricultural Progress*. National Planning Association, 1987b.

Meredith, WR; Bridge. Jr RR. "Genetic Contributions to Yield Changes in Upland Cotton." In Fehr, W.R; ed; *Genetic Contributions to Yield Gains of Five Major Crop Plants*. Madison, WI: CSSA, 1984, 75-87.

Merrill Lynch. *Ag. Biotech Update: Franken-Potato Revisted.* Merrill Lynch, Global Securities Research & Economics Group, Global Fundamental Equity Research Department, February 1, 1999.

_________. *Global Chemical Benchmark: The New Ag Products Industry, Chemicals.* Merrill Lynch, Global Securities Research & Economics Group, Global Fundamental Equity Research Department, July 7, 2000.

Monsanto Company. "About Monsanto." Accessed at URL: http://www. monsanto.com/monsanto/ about/past/index5.htm, June 2, 2000.

Morris, ML; ed. *Maize Seed Industries in Developing Countries*. Boulder, CO: CIMMYT, 1998.

Morse, L. "Dow Chemical Buys Out Partner." *Financial Times*, May 16, 1997, p. 33.

Mycogen Corporation. "Dow Subsidiary Purchases Mycogen Shares in Tender Offer." October 5, 1998. Accessed at URL: http://www.mycogen.com/ pressrelease.asp?ID=11, August 24, 2001.

_________. "Dow to Acquire All Outstanding Mycogen Shares." August 31, 1998. Accessed at URL: http://www.mycogen.com/ pressrelease.asp? ID=14, August 24, 2001.

National Research Council, Board on Agriculture and Natural Resources, Committee on Genetically Modified Pest-Protected Plants. *Genetically Modified Pest-Protected Plants: Science and Regulation*. Washington, DC: National Academy Press, 2000.

Northrup-King Co. "Northrup-King Co. History." Accessed at URL: http://www.nk.com/helper/ 1998/12_28_history_northrupking.html, June 9, 2000.

Ollinger, M; Fernandez-Cornejo, J. *Regulation, Innovation, and Market Structure in the U.S. Pesticide Industry*. Agricultural Economic Report 719. U.S. Department of Agriculture, EconomicResearch Service, 1995.

Panos. "Greed of Need? Genetically Modified Crops." Panos Media Briefing No. 30. October 1998. Accessed at URL: http://www.oneworld.org/panos/

Pharmacia Corporation. "Corporate Facts." Accessed at URL: http://www.pharmacia.com/facts_monsanto.html, June 6, 2000.

Pioneer Hi-Bred International, Inc. "75 Years: A Legacy of Values." Accessed at URL: http://pioneer.com/ 75th/default.htm, August 24, 2001.

Plant Variety Protection Act. 7 U.S.C. § 2321-2583. 1970.

Pollak, A. "Green Revolution Yields to the Bottom Line." *The New York Times*, May 15, 2001.

Preckel, PV; Baker, TG; Eide. J. "Seed Corn Tournament and Excess Nitrogen Applications." Unpublished manuscript, Purdue University, February 12, 1997.

Newswire, PR. "Monsanto Completes Sale of Stoneville to Hicks, Muse, Tate & Furst and Emergent Genetics, Inc." December 29, 1999.

Newswire, PR; FirstCall (via COMTEX). "As Pharmacia Completes Monsanto Spinoff, Monsanto CEO Says Company Retains Focus and Strategic Direction, August 13, 2002." Accessed at URL: http://www.corporate-ir.net/ireye/ir_site.zhtml?ticker=MON&script= 410&layout=-6&item_id =325145 Sept 27, 2002.

Rivlin, R. "Monsanto Sells Seeds Business." *Financial Times*, August 7, 1999, p. 19.

Ross, D. "Antitrust Enforcement and Agriculture." Address before the Organization of Competitive Markets. U.S. Department of Justice, Nashville, TN, July 20, 2001, http://www.usdoj.gov/atr/public/speeches/8661.htm.

Ruttan, VW. "Biotechnology and Agriculture: A Skeptical Perspective." *AgBioForum* 2(1)(2000), 54-60. Accessed at URL: http://www.agbioforum.org/ vol2no1/ruttan.html.

Schimmelpfennig, DS; Thirtle, C. "Research Spillovers Between the European Community and the United States." *Contemporary Economic Policy*, Vol. 19, No. 4, October 1999, 457-68.

Schimmelpfennig, DS; Thirtle, C van Zyl, J. "Short and Long Run Returns to Agricultural R&D in South Africa, or Will the Real Rate of Return Please Stand Up?" *Agricultural Economics*, Vol. 24, No. 1, June 2000, pp.1-15.

Schmalensee, R; Willig, RD. *Handbook of Industrial Organization* Vol. II. In Arrow, K.J M.D. Intriligator. *Handbooks in Economics*. Amersterdam: Elsevier, 1989.

Schor, J. *The Evolution and Development of Biotechnology: A Revolutionary Force in American Agriculture*. AGES 9424. U.S. Department of Agriculture, Economic Research Service, August 1994.

Schroeter, JR. 1988. "Estimating the Degree of Market Power in the Beef Packing Industry." *Review of Economics and Statistics* 70(1988), 158-62.

Schroeter, JR; Azzam, A. "Marketing Margins, Market Power, and Price Uncertainty." *American Journal of Agricultural Economics* (1991), 990-99.

Sears, ER. "Genetics and Farming." In *Science of Farming - Yearbook of Agriculture 1943-1947*. U.S. Department of Agriculture, 1947, 245-55.

Seedquest. "AstraZeneca Merger Approved; U.S. Names Senior Management." Accessed at URL: http://www.seedquest.com/News/releases/europe/Zen eca/N1710.htm, April 6, 1999.

________. "Dupont and Pioneer Hi-Bred International, Inc. Merger Completed." Accessed at URL: *http://www.seedquest.com/News/releases/usa/ Dupont/n2145.htm*, October 1, 1999.

________. "KWS and Limagrain Merge Their Corn Activites in North America." Accessed at URL: http://www.seedquest.com/News/releases/europe/ Limagrain/n2406.htm, January 19, 2000.

________ "Monsanto Agrees To Purchase Cargill International Seed Operations in Central and Latin America, Europe, Asia, Africa." Accessed at URL: http://www.seedquest.com/News/Press%20releases/Monsanto/N1119.htm, June 29, 1998.

________. "Monsanto and Pharmacia & Upjohn To Merge, Creating Global Leader in Pharmaceuticals With Top-Tier Growth Prospects – Leading Global Agricultural Business To Be Operated as a Separate Entity, With a Partial IPO." Accessed at URL: http://www.seedquest.com/News/releases/usa/ Monsanto/n2331.htm, December 19, 1999.

Seed World. "Building Business With Seed Treatment," December 1999.

Sexton, RJ. "Imperfect Competition in Agricultural Markets and the Role of Cooperatives: A Spatial Analysis." *American Journal of Agricultural Eco nomics* (1990), 609-720.

Shields, M. "Novartis and AstraZeneca Mix Agribusiness." *American Seed Trade Association Newsletter*, December 1999. Accessed at URL: http://www.amseed.com/newsletter/newsletr.htm, January 19, 2000.

Shimoda, SM. "Agricultural Biotechnology: Master of the Universe?" *AgBioForum* 1(2) (1998), 62-68. Accessed at URL: http://www.agbioforum.missouri.edu, September 22, 1999.

Shoemaker, R; Harwood, J; Day-Rubenstein, K; Dunahay, T; Heisey, P; Hoffman, L; Klotz-Ingram, C; Lin, W; Mitchell, L; McBride, W; Fernandez-Cornejo, J. *Economic Issues in Agricultural Biotechnology*. Agriculture Information Bulletin 762. U.S. Department of Agriculture, Economic Research Service, February 2001.

Smale, M; Day-Rubenstein, K. "The Demand for Crop Genetic Resources: International Use of the U.S. National Germplasm System." *World Development* (2002), 1639-55.

Sprague, GF. "Hybrid Corn." In *After a Hundred Years - Yearbook of Agriculture, 1962*. U.S. Department of Agriculture, 1962, 106-107.

St. Louis Post-Dispatch. "Justice Officials Looking Into Flurry of Cottonseed Mergers." July 20, 1996: p. 1C.

Stoneville Pedigreed Seed Company. "News." Accessed at URL: http://www.stoneville.com/ news.htm, June 8, 2000.

Storck, WJ. "Pesticide Growth Slows." *Chemical and Engineering News*, November 16, 1987.

Strachan, JL. Private communication. U.S. Department of Agriculture, Agricultural Marketing Service, Plant Variety Protection Office, 2003.

Strachan, JM. *Plant Variety Protection: An Alternative to Patents*. U.S. Department of Agriculture, Agricultural Marketing Service, Plant Variety Protection Office, 1992.

Syngenta About Syngenta, Facts and Figures. Accessed at URL: http://www.syngenta.com/en/about_syngenta/figures.aspx, May 2003

Teitelbaum, GL; Figueroa. A. "Seed Companies, A Review of the Investment Case," Merrill Lynch, 1997.

Tripp, R. "Regulatory Issues: Varietal Registration and Seed Quality Control." In Morris, M.L; *Maize Seed Industries in Developing Countries*. Boulder, CO: CIMMYT, 1998, 159-73.

U.S. Congress, Office of Technology Assessment. *New Developments in Biotechnology Patenting Life— Special Report*. OTA-BA-370, April 1989.

U.S. Congress (USC). *Plant Variety Protection Act*. 7

U.S.C. 1970. § 2321-U.S. Department of Agriculture, Agricultural Marketing Service, Cotton Program, Market News Branch. *Cotton VarietiesPlanted, Crops of (1970-1997)*. Accessed at URL:http://www.ams.usda.gov/ cotton/cnpubs/htm, June 14, 2000.

U.S. Department of Agriculture. Agricultural Biotechology. Frequently Asked Questions. What Is Biotechnology? Accessed at URL: http://www. usda.gov/ agencies/biotech/faq/html, 2003

U.S. Department of Agriculture, Agricultural Marketing Service, Plant Variety Protection Office. "Science and Technology Programs." Accessed at URL: http://www.ams.usda.gov/science/PVPO/pvp.htm, August 10, 2000.

U.S. Department of Agriculture, Agricultural Marketing Service. *Plant Variety Protection Act and Regulations and Rules of Practice*. Revised March 2001, 45 pp, http://www.ams.usda.gov/science/ PVPO/PVPO_ Act/whole.pdf

U.S. Department of Agriculture, Agricultural Marketing Service, Plant Variety Protection Office. "Plant Variety Protection Listings." Accessed at URL: http://www.ars-grin.gov/cgi-bin/npgs/html/ pvplist.pl? May 30, 2001a.

U.S. Department of Agriculture, Agricultural Marketing Service, Plant Variety Protection Office. *Cotton PVP Accessions*. Accessed at URL: http://www. arsgrin.gov/cgi-bin/npgs/html/pvp.pl?Cotton, May 30, 2001b.

U.S. Department of Agriculture, Animal and Plant Health Inspection Service. "Regulating Biotechnology in Agriculture." Accessed at URL: http:// www.aphis.usda.gov/oa/biotech.html, August 14, 2000.

U.S. Department of Agriculture, Cooperative State Research, Education, and Extension Service, Current Research Information System. *Manual of Clas sification of Agricultural and Forestry Research: Classifications used in the Current Research Information System, Revision V*. 1993.

_________, Current Research Information System, *Inventory of Agricultural Research*. 1968-1996.

_________, Information Systems for Biotechnology. "Field Tests in the U.S; 1988-2000." Accessed at URL: http://www.nbiap.vt.edu/cfdocs/ fieldtests1.cfm, July 3, 2000b.

_________, Information Systems for Biotechnology. "Background on the Environmental Release Database." Accessed at URL: http://www.nbiap.vt.edu/ biomon/explain.htm, August 14, 2000b.

U.S. Department of Agriculture, Economic Research Service. *Agricultural Resources Inputs Situation and Outlook*, AR-13, February 1989.

_________. *Agricultural Resources Inputs Situation and Outlook*, AR-32, October 1993.

_________. *Agricultural Resources and Environmental Indicators, 1996-97*. Agricultural Handbook 712. July 1997.

_________. *AREI Updates 1998*, No. 8, December 1998.

________. "*Farm Sector Performance – Situation and Outlook.*" Accessed at URL: http://www.ers.usda.gov/ briefing/farmincome/s&o.htm, August 20, 1999.

________. *Agricultural Resources and Environmental Indicators 2000*, AREI 2000, August 2000.

________. *Agricultural Resource Management Survey (ARMS), Recommended Data.* Accessed at URL:http://www. ers.usda.gov/ Briefing/ARMS/ moredata.htm, March 28, 2002.

U.S. Department of Agriculture, National Agricultural Statistics Service. *Agricultural Statistics.* Various years.

U.S. Department of Agriculture, National Agricultural Statistics Service. "Published Estimates Data Base." Accessed at URL: www.nass.usda. gov:81/ipedb/, accessed May 25, 2001.

U.S. Department of Energy, Energy Information Agency. "Electricity." Accessed at URL: www.eia.doe.gov/fuelelectric.html, May 21, 2001.

U.S. Department of Labor, Bureau of Labor Statistics. *Bureau of Labor Statistics Data: Series ID: EES00500049, Seasonally Adjusted, Industry: Total Private Industry, SIC Code: na, Average Hourly Earnings, 1982 Dollars; 2: Series ID: EES30000006, Seasonally Adjusted, Industry: Manufacturing, SIC Code: 20-39, Average Hourly Earnings of Production Workers.* Accessed at: URL: www.bls.gov, May 18, 2001

U.S. International Trade Commission. *Industry & Trade Summary: Seeds.* USITC 2859, May 1995.

Van Wijk, Jeroen, Cohen, JI; Komen. J. *Intellectual Property Rights for Agricultural Biotechnology: Options and Implications for Developing Countries. A Biotechnology Research Management Study.* ISNAR Research Report No. 3. The Hague: International Service for National Agricultural Research, 1993.

Virginia Polytechnic Institute and State University. "Information Systems for Biotechnology, Environmental Release Database of Field Test Releases in the U.S." Accessed at URL: www.isb.vt.edu/ cfdocs/fieldtests1.cfm, June 28, 2001.

The Wall Street Journal Interactive Edition. "Monsanto To Expand Internationally With $1.4 Billion Purchase From Cargill," June 29, 1998. Accessed at URL: http://interactive.wsj.com, June 29, 1998.

Wann, JJ; Sexton, RJ. "Imperfect Competition in Multiproduct Food Industries With Application to Pear Processing." *American Journal of Agricultural Economics* (1992), 980-90.

Williamson, OE. "Economies as an Antitrust Defense: The Welfare Tradeoff. *American Economic Review* (58)(1968),18-36.

Zeneca Agrochemicals. "Plant Biotechnology." Accessed at URL: http://www.zeneca.com/page.asp? section=research&sub=biotech, June 6, 2000.

End Notes

[1] This genetic malleability of corn, on the other hand, meant that farmers could more easily select for characteristics they wanted on their own.

[2] Acronym from the French *Union Internationale pour la Protection des Obtentions Végétales*.

[3] This does not apply to open-pollinated corn because it would not be "genetically uniform and stable through successive generations."

[4] This case is particularly relevant to the issue of genetically engineered nonhybrids such as herbicide tolerant "Roundup Ready" soybeans. Monsanto, the largest producer of these varieties, required farmers purchasing the soybean seed to enter into contracts that specifically prevented them from saving seed for future planting.

[5] For some quantitative results, see Smale and Day-Rubenstein (2002).

[6] UPOV is an intergovernmental organization headquartered in Geneva, Switzerland. UPOV was established by the International Convention for the Protection of New Varieties of Plants with the objective of protecting new varieties of plants through intellectual property rights. The Convention was adopted in Paris in 1961, and it was revised in 1972, 1978, and 1991. These revisions, or amendments, are referred to as the "1972 Act," "1978 Act," and "1991 Act" (UPOV, 2003).

[7] Apart from corn, the only other field crops that have been successfully hybridized are sorghum and sunflower. Breeders also successfully hybridized a number of vegetable crops, such as onions, tomatoes, broccoli, cabbage, melons, and spinach, but the market shares of these crops are marginal, compared with those of corn (McMullen, 1987, p. 89; Leibenluft, 1981, p. 95). Until recently, farmers growing other major field crops remained dependent on saved seed, thus limiting the growth of seed industries for those crops: only 55 percent of the soybean acreage, 50 percent of cotton acreage, and 10 percent of wheat acreage was cultivated with purchased seed as late as 1982 (McMullen, 1987, pp. 86-87).

[8] The chemical industry experienced its most marked growth—15 percent or more annually—during the late 1960s and 1970s. Since then, market growth has been under 10 percent and was predicted to slow in the late 1990s (Storck, 1987).

[9] For comparison, the market structure of the seed industry is much more concentrated than the market for pesticides, another key agricultural input (Ollinger and Fernandez-Cornejo, 1995). In the seed and pesticide markets for corn and cotton, where both inputs are primarily purchased from the private sector, the CR4 ratio is higher in seed markets than in pesticide markets. From 1972 to 1989, the estimated CR4 ratio for the pesticide market averaged 45 percent, compared with 60 percent for the cotton seed market and 58 percent in the corn seed market, both of which have increased in recent periods. Recent data for corn, soybeans, and cotton, which represent a large share of the market, suggest that the seed industry is more concentrated than the pesticide industry.

[10] For example, it was reported that in 1999, after learning that the U.S. Department of Justice intended to sue over concerns about the anticompetitive effects in the cotton seed market, Monsanto abandoned its proposed acquisition of Delta & Pine Land Co., which could have combined the Nation's two largest cottonseed firms (Ross, 2001).

[11] It is worth noting that that the increasing proportion of expenditure on agricultural chemicals greatly exceeds the increase in plant breeding, as shown in table 25. This is attributed to the fact that research in agricultural chemicals has long been dominated by the private sector, while plant breeding was traditionally the domain of public sector investment and only beginning to attract private investment during the earlier years of this period (Heisey, 1999, p. 19).

[12] We approximate public sector spending on plant breeding with USDA and SAES expenditures on improving biological efficiency for fruits and vegetables and field crops. Improvement of biological efficiency research is described in research problem areas 304 and 307 of the *Manual of Classification of Agricultural and Forestry Research*, as research on "the ability of agriculture to meet the feed, food, and fiber needs of the American people and provide vital amounts of these commodities for exports" (USDA, 1993, p. 71). The specific areas of research contained in this category of the *Manual* relevant to plant breeding are (i) the identification of superior germplasm and breeding and selection of improved varieties, and (ii) the genetic and biological determinants of biological efficiency. These research areas are adequate to examine trends in public sector expenditure on plant breeding, and to make comparisons with private sector expenditures.

[13] Amount of time depends on availability of source germplasm. Time can also be reduced when breeders can use more than one cycle per year (e.g; through Southern Hemisphere shuttle breeding).

[14] Exante, two important factors suggested a profitable market, justifying the time and expense of improving seed through biotechnology R&D. First, the seed market for U.S. field crops is very large. Second, genetically engineered seeds, selling at a price premium over conventional seeds, are not substantially more expensive to produce after R&D and regulatory approval are complete.

[15] Another measure of research output is the number of patents, which can also provide an indication of concentration of research assets (Brennan et al; 2000). Considering the two major crops, corn and soybeans, the three top firms, DuPont/Pioneer, Monsanto/DeKalb and Novartis/Syngenta, held 46 percent of the biotech patents or patent applications as of 1996-97 (table 51).

[16] Note, however, that few firms dominated cotton field release approvals by APHIS during the first 5 years (1988-92). Hence, the four-firm concentration ratio during this period was 100 percent. New firms entered the market and pursued R&D in subsequent years.

INDEX

#

20th century, 4, 85, 86, 87, 88, 91, 92, 138, 199

A

absorption, 67
access, 3, 6, 7, 13, 20, 21, 22, 23, 25, 26, 27, 35, 36, 40, 56, 59, 61, 62, 65, 66, 127, 133, 134, 141, 145
accessibility, 56
accessions, 8, 14, 16, 19, 23, 31, 46, 48, 59, 60, 83
accounting, 42, 43, 82, 88, 199
acquisitions, 19, 85, 86, 87, 137, 139, 141, 151, 152, 156, 189, 222, 224
adaptability, 2
adaptation, vii, 1, 2, 3, 4, 10, 11, 18, 19, 20, 23, 24, 25
Africa, 26, 27, 54, 121, 123, 235
age, 86
agencies, 83, 127, 133, 135, 137, 144, 170, 237
agricultural producers, 5, 125
Agricultural Research Service, 7
agricultural sector, 1, 91, 125, 128
agricultural seeds, vii, 86
agriculture, 31, 38, 41, 51, 52, 66, 67, 89, 90, 91, 92, 103, 128, 129, 136, 138, 170, 199, 240
alfalfa, 11, 58
analytical framework, 17
ancestors, 71, 83
annual rate, 26, 98, 178
anther, 93
Argentina, 87, 104, 116, 120, 122
arithmetic, 90
Armenia, 104
ARS, 23, 31, 166, 196, 209, 214, 216, 217
Asia, 12, 26, 27, 31, 54, 58, 73, 117, 121, 123, 235
ASS, 109, 111, 113, 115
assessment, 5, 8, 20, 35, 41, 72
assets, 240
ASTA, 130
asymmetry, 62
authorities, 136
authority, 135, 136
authors, 43, 44
aversion, 171
avoidance, 82
Azerbaijan, 104

B

bacteria, 7, 12, 94, 130
bacterium, 96, 131

banks, 16, 23, 36, 37, 41, 42, 47, 48, 54, 56, 59, 60, 64, 66, 67, 69, 82, 83
barriers, 222
barriers to entry, 222
base, 6, 26, 27, 42, 50, 53, 67, 70, 71
basic research, 39, 47
BD, 225, 226
Belarus, 104
Belgium, 155
beneficiaries, 46
benefits, vii, 1, 2, 3, 5, 14, 16, 17, 18, 19, 21, 22, 25, 31, 33, 34, 35, 37, 41, 42, 43, 44, 45, 46, 47, 48, 49, 51, 54, 56, 58, 60, 61, 62, 64, 65, 66, 75, 77, 125, 129, 146, 169, 171, 197
biochemistry, 131
biodiversity, 21, 28, 47, 51, 53, 65, 73, 74
biological processes, 126, 133
biosafety, 65
biotechnology, 16, 65, 85, 86, 89, 90, 94, 126, 127, 130, 131, 133, 136, 137, 139, 141, 142, 153, 190, 191, 194, 199, 200, 219, 222, 240
biotic, 9, 10, 11, 12, 19, 20, 24, 27, 54
bounds, 189
Brazil, 12, 104, 120, 154
breakdown, 189, 207
Bureau of Labor Statistics, 238
businesses, 85, 86, 88, 137, 138, 139, 141, 152, 155

C

cabbage, 95, 239
calcium, 12
capacity building, 66
carbon, 53
case study, 28
cash, 169
CBD, 29, 65, 66
Central Asia, 31
certificate, 201
certification, 135, 137
challenges, 77
changing environment, vii, 1, 58
chemical(s), 43, 70, 90, 126, 130, 139, 140, 142, 152, 153, 156, 172, 179, 180, 181, 182, 183, 184, 185, 239, 240
chemical industry, 239
Chile, 87, 117, 120, 122
China, 54, 58, 81, 87, 95, 102, 104, 123
chromosomal abnormalities, 39
civil action, 129, 130
classes, 68
cleaning, 139, 145, 157
climate, vii, 1, 2, 3, 4, 5, 7, 9, 10, 11, 13, 17, 18, 19, 20, 23, 24, 25, 29, 32
climate change, vii, 1, 2, 3, 4, 5, 7, 9, 10, 11, 13, 17, 18, 19, 20, 23, 24, 25, 29, 32
coffee, 64
collaboration, 170
colleges, 137, 144, 170
Colombia, 120
commercial, 2, 3, 8, 13, 17, 18, 19, 20, 21, 22, 32, 62, 63, 64, 66, 87, 88, 91, 92, 95, 102, 104, 137, 138, 142, 186, 200
commercial crop, 3, 62
commodity, 68, 146, 173, 174, 175
communication, 79, 80, 147, 230, 232, 236
community, 8, 67
compensation, 13, 125
competition, 142, 144, 222
competitive advantage, 171
competitiveness, 145, 189
competitors, 125, 138, 222
complement, 9, 194
complexity, 20
compliance, 134, 172
composition, vii, 27, 86, 126, 129
compounds, 66
conditioning, 85, 86, 143, 144, 145, 146
confinement, 136
conflict, 25
conformity, 128, 130
Congress, 32, 78, 126, 128, 236
consensus, 5, 34, 76
consent, 21
conservation, 2, 3, 5, 7, 9, 14, 19, 21, 22, 25, 29, 31, 33, 34, 35, 36, 37, 40, 41, 46, 47,

48, 49, 51, 57, 58, 59, 60, 61, 62, 64, 65, 66, 67, 68, 69, 78, 133
conserving, 14, 21, 35, 42, 56, 58, 59, 61, 64, 67
consolidation, 200
Constitution, 32, 128
consumers, 42, 44, 46, 62, 136, 169, 171, 197
consumption, 51, 54, 106, 197
controversial, 62
convention, 126, 133, 134
Convention on Biological Diversity(CBD), 21, 22, 29, 32, 36, 65, 79, 134
cooperation, 21
Cooperative State Research, Education, and Extension Service, 237
coordination, 21, 41, 61, 65, 127, 133
cost, 5, 7, 18, 19, 35, 43, 46, 51, 59, 82, 88, 103, 106, 111, 112, 113, 114, 115, 116, 125, 141, 142, 146, 151, 168, 194, 196, 197
Court of Appeals, 128, 132
covering, 70, 126, 130, 168
critical value, 36
crop genetic resources, vii, 1, 2, 3, 5, 20, 21, 23, 25, 32, 33, 35, 40, 41, 49, 50, 59, 60, 64, 67, 74
crop producers, 44
crop production, 19, 34, 43, 73, 103
crop productivity, vii, 1, 2, 4
crop varieties, 2, 3, 5, 7, 9, 24, 34, 38, 41, 53, 59, 62, 64, 69, 70, 90, 193, 197, 200, 201, 219
crop yields, 4, 34, 35, 36, 56, 82, 85, 90, 95, 98, 101, 138
crude oil, 131
CSREES, 177, 178
CT, 75
cultivars, 3, 6, 26, 27, 39, 47, 97, 200
cultivation, 102
culture, 95, 131
cyst, 19

D

danger, 59
database, 16, 31
deflator, 113, 115, 140
degradation, 51
Delta, 141, 153, 155, 157, 162, 165, 166, 214, 217, 218, 239
Denmark, 104, 122
Department of Agriculture, 28, 29, 72, 75, 76, 80, 81, 136, 170
Department of Energy, 238
Department of Justice, 234, 239
depression, 93
deregulation, 89, 219, 221
developed countries, 35, 44, 51, 62, 68
developing countries, 26, 27, 35, 36, 38, 43, 51, 54, 55, 60, 62, 63, 64, 68, 82, 95, 127, 133, 134
diffusion, 138
diminishing returns, 17
direct measure, 71
direct payment, 68
diseases, 2, 3, 5, 9, 10, 12, 33, 34, 35, 38, 40, 41, 48, 49, 56, 71, 82
displacement, 55, 61
distinctness, 131
distribution, 4, 10, 15, 23, 44, 48, 49, 50, 52, 69, 71, 75, 141, 143, 144, 145, 146, 147, 151, 153, 193, 197
divergence, 59
diversity, 5, 6, 7, 26, 29, 33, 34, 35, 36, 37, 38, 39, 40, 49, 50, 51, 52, 53, 54, 55, 56, 57, 58, 59, 60, 61, 62, 63, 65, 67, 69, 70, 71, 83, 133, 134
DNA, 3, 6, 7, 70, 95, 131, 199, 209, 221
domestication, 54
dominance, 34, 193
donors, 31
double helix, 95
drainage, 82
drought, 2, 3, 5, 10, 12, 18, 20, 38, 47, 96, 101
drugs, 21, 155
drying, 145

E

earnings, 62
Eastern Europe, 121, 122
ecological systems, 38
economic assessment, 35
economic incentives, 13, 27, 38
economic methodology, 37
Economic Research Service, 1, 6, 8, 10, 11, 18, 28, 29, 33, 72, 74, 75, 76, 81, 85, 228, 229, 230, 232, 235, 236, 237
economic values, 51
economic welfare, 35, 44
economics, 4, 5, 19, 46, 47, 48
economies of scale, 139, 142, 144, 172, 197
ecosystem, 38
educators, 7
Egypt, 31
empirical studies, 19, 46
endosperm, 91
enforcement, 63
engineering, 12, 52, 90, 91, 94, 126, 130, 131, 170
environment(s), 4, 9, 12, 26, 38, 39, 54, 59, 65, 70, 89, 101, 136, 194, 202
environmental conditions, vii, 1, 34, 37, 42, 58
environmental factors, 52
Environmental Protection Agency, 137
environmental services, 53
environmental stress(es), 5, 11, 96
enzyme, 70
EPA, 137
equilibrium, 44
equipment, 126, 130, 146, 197
erosion, 49, 56
ERS, 4, 53, 103, 133, 168, 227
Europe, 27, 82, 117, 120, 121, 122, 231, 235
European Community, 234
evidence, 2, 10, 63, 98, 131, 186
evolution, vii, 86, 147
exclusion, 13, 129
expenditures, 16, 31, 42, 63, 87, 88, 102, 105, 107, 108, 109, 141, 169, 172, 176, 177, 178, 186, 187, 189, 190, 192, 193, 194, 240
expertise, 64, 67
experts, 61
exploitation, 5
exporter, 87, 116
exports, 87, 116, 117, 118, 120, 124, 240
externalities, 169, 171

F

families, 68
farming techniques, 51
farms, 145
FAS, 119, 121, 123, 124
FDA, 137
Federal Government, 144
Federal Register, 137
fertility, 11, 101
fertilization, 93
fertilizers, 43, 82, 95, 97, 99
fiber, 41, 96, 196, 240
field crops, 8, 11, 86, 88, 89, 96, 97, 102, 103, 106, 112, 141, 142, 147, 151, 187, 189, 193, 195, 198, 199, 201, 203, 204, 222, 239, 240
field tests, 89, 135, 136
financial, 17, 22, 25, 36, 43, 64, 83, 88, 126, 128, 137, 147, 171
financial resources, 88, 137
financial support, 43
firm size, 196
fish, 68
flexibility, 12
flowers, 116, 117
food, 22, 35, 36, 41, 44, 60, 62, 65, 66, 68, 90, 95, 96, 119, 133, 137, 139, 142, 153, 156, 172, 240
Food and Drug Administration, 137
food prices, 44, 62
food production, 90
food security, 22, 95, 133
forage crops, 116
force, 21, 22, 38, 48, 65, 66, 138, 199
formation, 52, 172

formula, 23, 146
foundations, 68, 127, 133
France, 104, 116, 117, 120, 122, 124, 126, 140
frequency distribution, 48
fruits, 60, 240
funding, 14, 19, 34, 41, 60, 61, 67, 127, 133, 170, 193
funds, 40, 60, 61, 64, 67, 68, 69
fungi, 101, 130

G

GAO, 53, 59, 60, 81, 83
GATT, 72, 134
GDP, 140
GDP deflator, 140
gene pool, 8, 16, 52
gene transfer, 6
General Accounting Office, 35, 37, 57, 81
General Agreement on Tariffs and Trade, 134
genes, 3, 6, 12, 13, 14, 20, 23, 24, 32, 38, 39, 48, 52, 64, 66, 70, 83, 94, 127, 132, 134
genetic diversity, 5, 6, 7, 26, 29, 33, 34, 35, 36, 37, 38, 49, 50, 51, 52, 53, 54, 55, 56, 57, 58, 60, 62, 63, 69, 70, 71, 133, 134
genetic drift, 83
genetic engineering, 12, 52, 90, 94, 126, 130, 131
genetic information, 13
genetic marker, 3, 20
genetic resource use, vii, 2, 3, 4, 5, 16, 17, 19, 25, 69
genetic traits, vii, 1, 3, 4, 13, 18, 64, 94
genetics, 16, 18, 24, 95, 142, 189
genome, 16
genomics, 5, 11, 32, 48
genotype, 95
geometry, 32
Georgia, 104, 106, 213, 217
Germany, 81, 104, 120, 122
germination, 135, 145
germplasm, 2, 6, 7, 8, 9, 10, 11, 12, 14, 15, 16, 17, 19, 21, 22, 23, 25, 26, 30, 31, 33, 34, 35, 36, 37, 39, 40, 41, 43, 46, 47, 55, 56, 57, 58, 60, 62, 63, 64, 65, 66, 68, 69, 82, 83, 133, 189, 194, 240
governments, 68, 127, 133
grants, 127, 130
graph, 32
grasses, 93
grasslands, 38
Greece, 120
Green Revolution, 12, 53, 73, 75, 79, 80, 85, 90, 95, 234
greening, 80
GRIN, 8, 31
grouping, 95
growth, 4, 12, 34, 37, 42, 43, 44, 51, 60, 82, 85, 86, 89, 90, 95, 97, 99, 101, 117, 138, 139, 142, 172, 178, 189, 198, 199, 202, 239
growth accounting, 42, 43
growth rate, 117
GSA, 162, 165
Guatemala, 31
guidelines, 21

H

habitat(s), 31, 34, 35, 38, 46, 50, 51, 52, 53, 55, 58, 59, 61
harmonization, 23
height, 17, 18, 70
herbicide, 89, 96, 139, 142, 202, 219, 239
heredity, 91
history, 87, 199, 233
Hong Kong, 121
horticultural crops, 189
host, 50
human, 54, 199
Hungary, 104
hybrid, 53, 85, 87, 90, 91, 92, 94, 95, 99, 101, 129, 131, 137, 138, 139, 151, 157, 169, 172, 189, 219
hybridization, 91

I

ID, 115, 233, 238
ideal, 101
identification, 10, 17, 49, 91, 240
identity, 59, 135
imitation, 125
impacts, 62, 63
imports, 87, 117, 118, 119, 122
improvements, 2, 13, 19, 25, 34, 35, 36, 37, 43, 44, 46, 75, 82, 86, 87, 96, 98, 99, 102, 126, 129, 134, 143, 171
inbreeding, 91, 93
income, 26, 27, 40, 44, 62, 67
India, 75, 95, 104, 123
individuals, 49, 51, 61, 126, 145
industrial organization, 70
industries, 88, 168, 239
industry, vii, 21, 42, 85, 86, 87, 88, 125, 128, 129, 137, 138, 139, 140, 141, 142, 143, 144, 147, 148, 151, 153, 156, 157, 160, 166, 168, 169, 172, 189, 200, 222, 239
INF, 29
inflation, 88
informed consent, 21
infrastructure, 141
ingredients, 142
inheritance, 91
innovator, 129
insects, 7, 9, 96
institutional change, 172
institutions, 4, 21, 23, 31, 61, 62, 63, 83, 89, 127, 133, 136, 156, 157, 159, 169, 170, 194, 202, 222
integration, 144
integrity, 57
intellectual property, 1, 3, 4, 14, 20, 23, 25, 36, 40, 41, 60, 61, 63, 64, 66, 87, 91, 126, 128, 132, 133, 134, 141, 144, 156, 224, 239
intellectual property rights, 20, 23, 25, 36, 41, 60, 61, 63, 64, 87, 126, 128, 133, 134, 144, 156, 224, 239
international investment, 61
international standards, 128, 130, 134
international trade, 51
International Treaty on Plant Genetic Resources for Food and Agriculture, 22, 34, 36, 38, 66
intervention, 54, 60
inventions, 23, 36, 60, 63, 65, 126, 128, 130, 134
inventors, 14, 32, 126, 128
invertebrates, 136
investment(s), 1, 5, 14, 16, 17, 21, 31, 35, 36, 40, 46, 47, 59, 60, 61, 82, 87, 125, 133, 138, 139, 142, 144, 169, 171, 172, 189, 193, 197, 224, 240
investors, 145, 171
Iowa, 76, 152, 154, 158, 209, 210, 212, 217, 228, 229, 230, 231
IPO, 235
IPR, 61, 62, 63, 65, 66, 67, 126, 127, 128, 129, 130, 131, 132, 133, 134, 135, 144
irrigation, 82, 116
issues, 9, 30, 31, 51, 66, 77, 126, 127, 129, 133, 134
Italy, 82, 87, 104, 116, 117, 120, 122, 124

J

Japan, 87, 102, 104, 116, 117, 121, 123, 124
joint ventures, 139, 156

K

Kazakhstan, 104
Kenya, 78
Korea, 121
Kyrgyzstan, 104

L

labeling, 135
land tenure, 51
Latin America, 54, 235
laws, 91, 125, 128, 130, 136
LDCs, 78

lead, 31, 90, 157
leadership, 140
learning, 239
legal protection, 87, 125
legislation, 22, 126, 128, 129, 134, 170
life sciences, 142, 151, 153
light, 2, 136, 199
livestock, 41, 197
Louisiana, 106
lower prices, 169

M

machinery, 43, 99, 126, 130, 172, 179, 180, 181, 182, 183, 184, 185
magnitude, 42, 48, 189, 197
maize, 34, 46, 54, 58, 66, 82
majority, 47, 49, 89, 199, 202, 224
management, 9, 37, 53, 67, 82, 98, 146, 177, 178
manufacturing, 140, 156
manufacturing companies, 140
mapping, 20
marginal costs, 18, 46, 171
market concentration, 142, 147, 148, 156, 222
market failure, 171
market share, 141, 147, 148, 149, 150, 153, 155, 158, 159, 166, 168, 239
market structure, 239
marketing, 85, 86, 102, 103, 106, 138, 143, 144, 145, 146, 147, 197, 219
marketplace, 131, 135
materials, 6, 7, 8, 10, 11, 13, 14, 16, 17, 19, 20, 22, 23, 25, 26, 27, 31, 32, 36, 47, 48, 52, 56, 59, 60, 62, 63, 64, 65, 66, 70, 83, 104, 127, 133, 134, 169, 196
matter, 95, 126, 129
measurement(s), 47, 49, 50
medical, 21
medicine, 197
Mediterranean, 82
membership, 133
mergers, 85, 86, 87, 137, 139, 141, 160, 189, 222, 224
methodology, 37, 42, 131
Mexico, 55, 58, 76, 77, 80, 87, 104, 116, 117, 120, 122, 124, 157, 161, 164, 232, 233
microorganism(s), 38, 51, 94, 127, 131, 136
Middle East, 117, 121
Minneapolis, 82
Missouri, 167
models, 10, 16, 20, 32, 46, 48
modernization, 91
modifications, 32
moisture, 9, 10, 106, 116, 135, 145
moisture content, 135, 145
Moldova, 104
molecular biology, 48, 49
Montana, 216, 217
motivation, 41
Multilateral, 22, 64, 66
multinational corporations, 139
multinational firms, 142, 143, 148
multiplication, 144
mutation(s), 6, 50, 91

N

National Academy of Sciences, 77
National Research Council, 5, 6, 30, 37, 49, 50, 51, 53, 55, 67, 77, 233
nationality, 151
natural disaster(s), 57
natural selection, 91
natural selection processes, 91
negotiating, 22
Netherlands, 87, 104, 116, 117, 122, 124, 155
New South Wales, 72
New Zealand, 68, 123
NGOs, 62
nitrogen, 99, 101, 146
North Africa, 26, 27, 54
North America, 116, 117, 158, 160, 167, 235
NRC, 30, 37, 60, 63, 77
nutrient, 96
nutrition, 94, 101, 153

O

obstacles, 1, 23
Oceania, 121, 123
OH, 113, 115
oil, 131, 152
Oklahoma, 161, 164, 167
operations, 142, 150
opportunities, 23, 88, 139, 142
organism, 6, 7, 89, 94, 136
organize, 57
overlap, 11, 26, 64
ownership, 37, 129, 153, 155, 156, 201, 224, 225

P

Pakistan, 76, 80
parentage, 71, 83
parents, 27
participants, 128
patents, 24, 62, 83, 127, 128, 129, 131, 132, 134, 135, 197, 224, 225, 240
pathogens, 9
pathways, 4
pedigree, 26, 56, 71, 83
permission, 23, 130, 134, 147
permit, 71, 135, 136, 202
Peru, 12, 31
pesticide, 137, 142, 143, 151, 153, 239
pests, 2, 3, 5, 9, 10, 12, 33, 34, 35, 38, 40, 41, 48, 49, 56, 71, 82, 101, 136, 143, 144
Petroleum, 140, 158
pharmaceutical(s), 21, 31, 47, 65, 82, 96, 139, 142, 150, 151, 153, 179, 180, 181, 182, 183, 184, 185
phenotype, 202
Philippines, 12, 31
physiology, 93
pipeline, 10
planning decisions, 60
plants, 5, 6, 12, 31, 36, 39, 54, 59, 83, 91, 93, 94, 95, 126, 127, 129, 131, 132, 133, 136, 198, 199, 239
Poland, 104
policy, 14, 36, 37, 59, 61, 64
policy instruments, 37
policymakers, 34, 40, 52, 56, 58
pollen, 93, 94
pollination, 93, 94, 128, 145
pollinators, 58
pools, 16, 52
population, 40, 44, 90
population growth, 90
positive externalities, 169
potato, 93
potential benefits, 1, 25, 47
poverty, 51
precipitation, 10, 20, 32, 106
preparation, 145, 198
present value, 43, 44, 46
preservation, 34, 35, 36, 41, 51, 53, 58, 65, 67, 68, 134
President, 79
price index, 107, 108, 109
pricing behavior, 145
private benefits, 48
private firms, 63, 139, 144, 145, 146, 147, 156, 157, 160, 172, 189, 194, 222
private investment, 5, 138, 172, 197, 240
private sector, 6, 13, 14, 16, 62, 87, 88, 89, 125, 133, 138, 139, 141, 144, 146, 147, 156, 157, 158, 159, 169, 171, 172, 186, 189, 192, 193, 194, 197, 199, 224, 239, 240
private sector investment, 133
probability, 17, 19, 48, 49
probability distribution, 48, 49
producers, 5, 42, 44, 61, 125, 157, 171, 197
product performance, 144
production costs, 146
production function, 43
productivity growth, 34, 37, 82
profit, 3, 5, 141, 145, 168
profit margin, 145
profitability, 63, 102
project, 10, 61
propagation, 94

property rights, 20, 23, 25, 36, 41, 60, 61, 63, 64, 65, 87, 125, 126, 127, 128, 129, 130, 131, 132, 133, 134, 135, 136, 139, 144, 147, 156, 172, 197
protection, 23, 24, 40, 62, 65, 87, 89, 95, 96, 125, 126, 127, 128, 129, 130, 131, 132, 133, 134, 135, 136, 152, 172, 197, 199, 201
proteins, 83
public domain, 88, 127, 134, 137, 139
public expenditures, 172, 177, 178
public goods, 13, 20, 32, 33, 35, 40, 61, 64, 69
public investment, 1, 20, 31, 36, 59, 60, 171
public policy, 14, 36
public resources, 59
public sector, 3, 5, 9, 13, 25, 40, 56, 63, 88, 138, 146, 156, 157, 159, 169, 170, 171, 172, 186, 189, 193, 194, 240
public support, 25
pure line, 189
purity, 135, 144, 145, 197
PVP, 89, 94, 95, 197, 199, 200, 201, 203, 204, 237

Q

quality assurance, 137
quality standards, 145

R

R&D investments, 142, 144, 171, 197
rainfall, vii, 1, 2, 5, 106
rate of change, 71
rate of return, 32, 43, 171, 186
ratification, 128, 134
RE, 230, 231
real terms, 14, 87, 102, 106, 172, 193
recommendations, 37
regenerate, 6, 7
regions of the world, 36
regulations, vii, 86, 125, 128, 136, 172
regulatory framework, 129, 136
relatives, 2, 3, 4, 6, 7, 8, 9, 11, 13, 16, 17, 23, 25, 26, 27, 31, 32, 34, 35, 37, 39, 40, 41, 46, 47, 48, 51, 53, 55, 57, 58, 60, 61, 62, 63, 69
relevance, 128, 133
replacement, 54
reproduction, 54, 91, 94, 125
requirements, 32, 41
resale, 132, 146
research facilities, 60, 166
research institutions, 83, 127, 133, 169
researchers, 7, 22, 24, 37, 68, 138, 169
reserves, 53, 61
resilience, 3, 13
resistance, 2, 7, 9, 10, 11, 12, 19, 20, 27, 35, 38, 39, 40, 41, 42, 47, 48, 50, 54, 56, 64, 71, 82, 89, 91, 94, 96, 125, 135, 142, 143, 144, 202, 208, 219, 220
resolution, 134
respect, 64, 66
response, 32, 194
restoration, 11
restrictions, 23, 24, 32, 58
revenue, 189, 190, 191
rewards, 1, 61, 125, 126, 128
rights, 20, 21, 23, 24, 25, 36, 41, 60, 61, 62, 63, 64, 65, 66, 87, 125, 126, 127, 128, 129, 130, 131, 132, 133, 134, 135, 136, 138, 139, 144, 146, 147, 153, 156, 172, 197
risk(s), vii, 1, 2, 4, 52, 56, 83, 89, 125, 135, 136, 144, 171, 202
risk aversion, 171
Romania, 121
root(s), 22, 60, 101
rowing, 31
royalty, 68
rules, 1, 3, 34
Russia, 83, 104

S

safety, 136
sanctions, 65
Saudi Arabia, 121

scale economies, 139, 146
science, 32, 91, 92, 95, 143, 151, 199, 237
scientific knowledge, 89, 125
scientific method, 91
scientific publications, 49
scientific understanding, 14
scope, 17, 127, 132, 142, 197
security, 22, 57, 95, 133
seed industry, vii, 85, 86, 87, 88, 125, 129, 137, 138, 139, 140, 142, 143, 144, 147, 148, 151, 153, 157, 160, 166, 168, 169, 172, 189, 200, 222, 239
seed markets, vii, 86, 102, 104, 117, 137, 138, 142, 146, 147, 239
seeding, 102, 103, 106, 116
seedlings, 126, 129
self-fertilization, 93
self-interest, 5
seller(s), 14, 139
Senate, 83, 134
sequencing, 20
services, 53, 68, 135
shape, 20
shareholders, 142
showing, 91, 197
shrubs, 118, 119, 130
SIC, 238
signs, 91
simulations, 47, 48
small businesses, 85, 88
small firms, 147, 148, 156
social benefits, 3, 25, 125
social class, 68
social consequences, 125
social marginal cost, 171
society, 1, 5, 14, 37, 47, 51, 168, 171
solution, 125
South Africa, 234
South America, 117, 120, 122
South Dakota, 106
South Korea, 121
Southeast Asia, 73
Soviet Union, 60, 82, 83, 104, 121, 122
soybeans, 8, 22, 32, 43, 44, 58, 82, 86, 88, 89, 95, 96, 97, 98, 99, 103, 106, 111, 113, 139, 141, 142, 147, 192, 194, 197, 199, 200, 202, 207, 212, 217, 219, 223, 224, 225, 239, 240
Spain, 104, 117, 120, 124
spatial information, 3
specialists, 8, 54
specialization, 62
speciation, 52
species, 6, 7, 22, 26, 31, 37, 38, 39, 47, 49, 50, 52, 53, 58, 59, 69, 70, 93, 94, 141
spending, 169, 171, 172, 186, 189, 240
spillovers, 169
Spring, 30, 77, 80, 114, 115
stability, 131
state(s), 27, 31, 32, 51, 65, 66, 126, 128, 129, 133, 134
statistics, 111, 229
sterile, 52
stigma, 93
stock, 144, 153
storage, 46, 57, 60, 147
strategic planning, 144
strategy, 68
stress, 2, 9, 10, 11, 12, 16, 19, 38, 41, 94
structure, vii, 32, 48, 86, 94, 131, 139, 143, 147, 151
substitutes, 58
Sudan, 32
sugarcane, 58
supervision, 145
supplier(s), 47, 63, 88, 117, 133, 137, 157
Supreme Court, 28, 62, 127, 131, 132
surplus, 87, 116, 137
survey, 62
survival, 7
Sustainable Development, 29
Sweden, 104
Switzerland, 127, 152, 239

T

Taiwan, 123
Tajikistan, 104
taxes, 68
technical assistance, 145

technical change, 3, 43
techniques, 6, 7, 17, 19, 24, 31, 38, 39, 48, 51, 94, 194, 200, 219
technological advances, 32, 131
technological change, 32, 43
technologies, 19, 42, 153
technology, 5, 17, 27, 64, 65, 66, 67, 93
technology transfer, 65, 67
temperature, 9, 20
tenure, 51
testing, 72, 89, 135, 136, 139, 200, 202, 224
texture, 91
Thailand, 123
threats, 34, 136
time lags, 20, 25, 197
tissue, 95
tobacco, 96
total factor productivity, 85, 90
trade, 31, 51, 87, 116, 117, 134, 160
trading partners, 117, 136
training, 64, 145
traits, vii, 1, 3, 4, 5, 6, 7, 10, 11, 12, 13, 14, 16, 17, 18, 19, 20, 23, 24, 25, 27, 31, 38, 39, 40, 46, 47, 48, 49, 52, 56, 64, 70, 89, 91, 93, 94, 95, 96, 127, 132, 143, 202, 208, 219
transaction costs, 63
transformation, 139, 156
transport, 52
transportation, 147
treatment, 146
trends, 38, 50, 56
tropical forests, 53
Turkey, 61
Turkmenistan, 104
turnover, 71

U

U.S. Department of Agriculture, 1, 28, 29, 33, 75, 76, 81, 85, 136, 170, 225, 226, 228, 229, 231, 232, 233, 234, 235, 236, 237, 238
U.S. Department of Labor, 238
U.S. policy, 34
UK, 73, 76, 78, 79, 155, 229
Ukraine, 104, 121
UN, 60, 134
underproduction, 171
uniform, 56, 129, 131, 133, 239
United Kingdom, 104, 120
United Nations, 21, 37, 72, 81, 134
United Nations Development Programme, 81
universities, 31, 137, 144, 170
urban, 51
urbanization, 40
Uruguay, 134
Uruguay Round, 134
USA, 75, 227, 231
Uzbekistan, 104

V

valuation, 5, 37, 42, 73
variables, 52
variations, 7, 38, 102
vegetables, 60, 93, 95, 116, 117, 199, 240
Venezuela, 120
vertical integration, 144
Vice President, 79, 229
vulnerability, 33, 34, 49, 55

W

Wales, 72
Washington, 28, 29, 30, 72, 73, 74, 75, 76, 77, 78, 79, 116, 226, 227, 231, 233
water, 12, 36, 101
wealth, 171
welfare, 35, 42, 44
well-being, 49
Western Europe, 117, 120, 122
wetlands, 38
wildlife, 53
Wisconsin, 227
working groups, 8
World Bank, 79, 81
World Trade Organization, 23, 65, 134

worldwide, 27, 34, 35, 53, 60, 67, 87, 90, 102, 133, 134, 139, 142, 153
WTO, 65, 134, 135

Y

Yale University, 75
yeast, 132
yield, 9, 10, 11, 12, 13, 21, 34, 35, 36, 40, 42, 43, 44, 46, 54, 60, 82, 86, 89, 90, 91, 95, 96, 97, 98, 99, 101, 116, 125, 129, 135, 144, 146